Starring the Text

Starring the Text

The Place of Rhetoric in Science Studies

Alan G. Gross

Southern Illinois University Press / *Carbondale*

Printed in the United States of America
18 17 16 15 5 4 3 2

Library of Congress Cataloging-in-Publication Data
Gross, Alan G.
Starring the text : the place of rhetoric in science studies / Alan G. Gross.
p. cm.
Includes bibliographical references and index.
1. Communication in science—Philosophy. 2. Rhetoric—Philosophy. I. Title.
Q223.G773 2006
501'.4—dc22 2005027884
ISBN-13: 978-0-8093-2695-2 (alk. paper)
ISBN-10: 0-8093-2695-7 (alk. paper)
ISBN-13: 978-0-8093-2696-9 (pbk. : alk. paper)
ISBN-10: 0-8093-2696-5 (pbk. : alk. paper)

The paper used in this publication meets the minimum requirements of American National Standard for Information Sciences—Permanence of Paper for Printed Library Materials, ANSI Z39.48-1992. ♾

To the memory of
my father, Jacob Gross,
and
my grandfather Jacob Jamron

Contents

Preface

Is rhetoric a master discipline, encompassing all others? Of course not. The claim that rhetoric is an intellectual tool useful in explaining the sciences rests on an alternate epistemological vision: an insistence that science is just one way of knowing. This claim in turn rests on a fundamental federalism about the domains of knowledge. In accordance with this federalism, rhetoric produces a knowledge different from that of science, different though not inferior: knowledge of science insofar as science is persuasive communication. Rhetoric "stars" the texts, tables, and visuals of science, that is, it makes their hermeneutic unraveling central. This is its role in science studies as one discipline among many joined in a common enterprise, a confederation of equally sovereign intellectual states: history of science, philosophy of science, and sociology of science.

This book is intended as a tribute to the durability and flexibility of the rhetorical tradition. In one sense, the application of the machinery of rhetoric to the texts of the sciences is merely an incremental move: scientific texts are, after all, just texts. But in another sense, such a move is analogous to Galileo's radical insistence that one physics be applied to the heavens and the earth. The texts of science are not just texts; they are what the Bible once was: revealed truth about the causal structure of the universe.

This book is a major refiguring of an earlier one, *The Rhetoric of Science.* I have changed the title of the original book because the new title is, in my view, an exact description of the current book's contents. In rewriting, I have also excised about half the material in the original book and have added a corresponding amount of material from my work in the intervening years. In addition, I have excised the radical epistemological inferences that were designed to provoke thought but that, as I might have anticipated, provided an easy target for the book's harshest critics. I have not altered my philosophical position, but I believe now that this position must be anchored in philosophical, not rhetorical arguments.

To one of the reviewers of this book, its case studies were bewildering in their variety. The purpose of this variety was not to confuse, of course, but to demonstrate that, in the sciences, wherever we turn, rhetorical analysis forms a profitable perspective for the critic. Still, I sympathize with the reviewer. In re-visioning the earlier work, I have tried to domesticate this variety. In part, I make explicit what was implicit: the deliberate choice of subject matter from the hard sciences and from both early and late modern science. My examples derive from biology and physics; they also contrast chronologically: seventeenth- and twentieth-century physics and nineteenth- and twentieth-century biology. Most importantly, I have restructured my cases to form an argument that seeks out the limits of rhetorical analysis, demonstrates the range of such analysis, and explores the place of this analysis among the other disciplinary endeavors that devote themselves to the study of the sciences.

This book is deeply indebted to the abundant scholarship in rhetoric of science. Although citations to this literature are few, this scarcity does not indicate lack of indebtedness; on the contrary, it indicates only that my debt to the literature is so deep and pervasive that isolated acknowledgments would misrepresent it. My work was formed in the climate of opinion of the three decades that I contributed to the field; throughout that time, the literature has constituted the scholarly air I breathed and still breathe.

I would like to thank Carol Berkenkotter, Randy Allen Harris, and Sara Newman for their considerable help with earlier drafts of this book. I would also like to thank John Angus Campbell and Jeanne Fahnestock, the two reviewers for Southern Illinois University Press, for their detailed and uniformly helpful criticisms. Finally, I would like to thank my wife, Suzanne Lee Gross, for more than a quarter-century of the emotional and intellectual support without which my work would have been difficult and, in some cases, impossible.

Part One

The Case for the Rhetorical Analysis of Science

1 The Achievements of Rhetoric of Science

In 1962 Max Perutz won the Nobel prize in chemistry; in 1995, in the *New York Review of Books*, in a review of Gerald Geisen's *Private Science of Louis Pasteur*, Perutz remarks in passing about the worth of rhetoric of science:

> According to Geisen, it is now commonplace among historians and sociologists of science that science, no less than any other form of culture, depends on rhetorical skills. I have known scientists who possessed great rhetorical skills which failed to conceal the shallowness of their research from their peers. On the other hand, Alexander Fleming's lectures put everyone to sleep, while his discovery of penicillin made him one of this century's most famous scientists. Good research needs no rhetoric, only clarity. The entire approach emphasizing "relative" truth seems to me a piece of humbug masquerading as an academic discipline; it pretends that its practitioners can set themselves up as judges over scientists whose science they fail to understand. (1995, 58)

In 2003, in her *Defending Science—Within Reason*, Susan Haak engages *The Rhetoric of Science* as follows:

> Gross's view of science is like an atheist's view of theology. A belief that a deity exists is the psychological anchor that makes a life in theology possible; however, an atheist would add, since this belief is false, theologians (though not exactly frauds, and not exactly writers of religious fiction) are in effect self-deceived fabulists. But science is different from theology in a crucial respect: that there is a real world which in a certain way and not other ways is a presupposition not only of scientific inquiry, but of all empirical inquiry, including the most ordinary everyday investigation into whether that check has cleared, what caused that leak in the roof, etc. (221)

These passages are interesting because they convey the remarkable news that a famous scientist and a respected philosopher of science have felt the need to notice rhetoric of science and, having noticed it, to ascend to polemic. I conjecture that there are two reasons for these two critics to be annoyed. The first, the existence of a substantial body of work in rhetoric of science by both rhetoricians and by their colleagues in philosophy, history, sociology, economics, sociology, linguistics, and library science, is the subject of this chapter; the second, the existence of a solid conceptual and methodological basis for a rhetoric of science, is the subject of chapter 2.

The Rhetoric of Science Policy

Science policy involves deliberative issues that fall readily within the traditional concerns of those trained in rhetorical analysis. Nevertheless, so strong was the traditional focus of the emerging discipline of speech communication on political oratory that the first rhetorical analysis of science policy was not made until 1953. In this study, Richard M. Weaver is concerned with an early climax in a continuing conflict in American public education, the place of evolution in the public-school biology curriculum. The focus of Weaver's study is the Scopes trial. In that trial, he concludes, the prosecution and the defense argued at cross purposes. The issue at hand was not the law against teaching evolution but the legality of Scopes's conduct under the law. Given this issue, the scientific testimony in favor of evolution was irrelevant. Indeed, even in the legislature, the question was not the truth of evolution but the right of the state to exclude from the curriculum what was for the people of Tennessee academic knowledge perhaps but religious heresy certainly.

More recent studies of policy issues relating to science—John Lyne on sociobiology, Arthur Walzer and myself on the *Challenger* disaster, Carolyn R. Miller on decision science, and Jeanne Fahnestock on science popularization—also exist entirely within the traditional bounds of classical rhetorical theory. Their novelty consists chiefly in their concern with the ethical issues science and technology leave in their wake and in their closer engagement with the science and technology that policy is designed to constrain. Lyne claims that the new science of sociobiology makes clear the need for a rhetoric of public policy. According to sociobiologists, biology is, to some large but undetermined extent, the cause of social behaviors: of friendship or fraud. But even if this were true, even if there were, for instance, genes for friendship or for fraud, the real question, the deliberative question, would be: what ought we to do about this fact? In Lyne's view, only rhetoric can address this question. Walzer and I claim that the cause of the *Challenger* disaster was not, as the presidential commission suggested and several communication scholars affirmed, faulty O-rings compounded by faulty communications. It was, rather, a cognitive and ethical failure, a failure of the deliberative process that should have led to a postponement of the launch.

Fahnestock and Miller also have strong views concerning the social and ethical implications of scientific and technological practices. In an article on decision science, Miller faults decision science for its misrepresentation of the way people actually make decisions. She has, she believes, two strong arguments in favor of her view. Decision science is intellectually flawed; moreover, despite its scholarly reputation, decision science is little used in actual practice. She suggests an alternate model for managerial deliberation, the ancient art of rhetoric. It is this ancient art and not its modern alternative that in her view best represents the way people actually deliberate about important issues.

Fahnestock is also concerned in her analyses with the social and ethical implications. In an examination of the rhetorical differences between science and its popularizations, Fahnestock makes the rhetorical point that science is written in the forensic genre, its popularizations in the epideictic. As a result, scientific claims in the popular media distort science, making its claims seem more certain than they actually are. In many cases, this is ethically indifferent. But not always. Fahnestock takes as her example a scientific paper that claims that boys may be better at mathematics than girls—genetically better. The original scientific claims are hedged and were soon disputed on various grounds. In the popular press, however, the claim was not at all hedged; moreover, there were no follow-up stories concerning its controversial nature.

The Rhetoric of Science Itself: The First Generation

The focus of rhetoricians on science itself, rather than science policy, represents a definite break with traditional rhetorical criticism. In this sense, all rhetoric of science is radical. Nevertheless, it is possible to analyze even the primary texts of science—for example, Galileo's *Two World Systems*, Isaac Newton's *Opticks*, and Charles Darwin's *Origin*—without opening the science itself to rhetorical scrutiny. These texts can be treated as communications designed to persuade relevant members of scientific communities concerning scientific claims that are already true, true as a result of a scientific rather than a rhetoric process. On the other hand, we can make the more radical claim that, in some sense, rhetoric is constitutive of scientific knowledge.

Four rhetoricians of science of the first generation—John Angus Campbell, Charles Bazerman, Greg Myers, and an earlier version of myself—do claim that rhetoric is, in some sense, constitutive of knowledge. Of the four, Campbell, whose discipline is speech communication, has been publishing on rhetoric of science for the longest time. Originally focused on the *Origin*, Campbell's work has since moved forward and backward in time, forward to Darwin on orchids, backward to Darwin's *Notebooks*. In all of his work, Campbell's message is the same: Darwin is the master rhetorician, willing even to distort and disguise his religious and scientific views if he believes that distortion and disguise will attain conviction on some issue central to evolutionary theory. Indeed, Darwin is a master rhetorician even in his *Notebooks*, whose audience is only Darwin himself. These are a testing ground for Darwin's theories, tested against the imaginary audience of such important potential objectors as his geological mentor and friend Charles Lyell. Of the relationship between the *Origin* and the *Notebooks*, there is, Campbell says:

> an unbroken continuity. . . . From his jotting in his first notebook through the sixth and final edition of the *Origin*, scientific discovery and rhetorical invention, technical and social reason, so effectively unite in Darwin's thought that one can only say that each is an aspect of a single logic of inquiry and presentation. (1990, 86)

While this conclusion shies away from implicating rhetoric in the content of science, it clearly asserts its integral relationship to scientific discovery.

Bazerman ranges more widely than Campbell: from Newton to Arthur Compton, from the style manual of the American Psychological Association to the reading habits of physicists. Unlike Campbell's, Bazerman's training is literary. The difference shows in his close, careful readings of his chosen texts, exegeses typical of literary criticism. These readings illustrate both successes and failures of communication. In his chapter on Newton, for example, Bazerman shows in great detail why the great scientist's first optical paper ultimately failed to convince. He also demonstrates how Newton learned from that failure to present essentially the same material in his later *Opticks* in a more convincing form. In his chapter on the APA style manual, we are the witnesses to the paradox of success. All scholarly papers in psychology are bound by the behaviorist strictures of the manual, although the field as a whole has left behaviorism far behind. In his conclusion, Bazerman gives a ringing endorsement of the constitutive nature of scientific rhetoric:

> Persuasion is at the heart of science, not at the unrespectable fringe. An intelligent rhetoric practiced within a serious, knowledgeable, committed research community is a serious method of truth seeking. The most serious scientific communication is not that which disowns persuasion, but which persuades in the deepest, most compelling manner, thereby sweeping aside more superficial arguments. Science has developed tools and tricks that make nature the strongest ally of persuasive argument, even while casting aside some of

> the more familiar and ancient tools and tricks of rhetoric as being only superficially and temporarily persuasive. (1988, 321)

But, as the title of his book *Shaping Written Knowledge: The Genre and Activity of the Experimental Article in Science* indicates, Bazerman sees rhetoric as shaping, not creating knowledge.

Myers expands the notion of text to include drafts as well as final versions; indeed, he sees rhetoric as an adjunct to and a deepening of sociolinguistic analysis. In one chapter, he follows two papers in biology through the process that leads to their final placement in a scientific journal. Initially, the claims in these papers exhibit the considerable breadth required of the prestigious general journals to which they are sent for review; but as rejection follows rejection, these broad claims are negotiated downward. Eventually, these papers are published in specialist journals with claims no wider than their specialist audiences will comfortably accept. The negotiating process is tied inextricably to persuasion and, consequently, to rhetoric; clearly, it is knowledge that is being negotiated or negotiated away:

> In one context, these cases suggest that the process of writing and revision of articles has an important consensus-building function. We have seen how this process maintains the homogeneity of the scientific literature. We have also seen how it shapes the research itself, Bloch, for instance, putting more and more emphasis on his data. In another context, that of literary criticism, these cases show the relations between texts, within the genre of the scientific article. The question in this context is not how reality is transformed in texts, but how it is made by texts. (1990, 98)

Myers shrinks from saying knowledge is rhetorical; presumably, he rests his rhetorical conservatism on the bedrock that it is the breadth, not the nature of the claims that is the object of contention.

On the question of the rhetorical constitution of knowledge, my *Rhetoric of Science* takes an explicitly radical position. In one chapter, for example, I compare James Watson and Francis Crick's famous paper on the structure of DNA with Watson's autobiography, *The Double Helix*, and find no interesting epistemological differences between the two:

> In the autobiographical case, persuasion is the product of a descriptive and processural presence, of facts shaped by an analogous adherence to a pattern supposed to underlie a life in science, a "theory" of such a life. In the scientific case, persuasion is the product of a descriptive and processural presence, of facts shaped by an analogous adherence to underlying stereochemical and genetic theories. (1996, 65)

But these epistemological inferences are grounded in a rhetorical analysis that does not depart from traditional roots. Commenting on the "immediately" in the last sentence of Watson and Crick's paper—"It has not escaped our notice that the specific pairing we have postulated immediately suggests a possible copying mechanism for the genetic material"—I say that this adverb is

> really an instruction to the reader to re-view the description and depiction of the DNA molecule, to see the dynamic possibilities of an entity hitherto viewed as static. We are asked to perceive a just-described static structure in new way; to undergo a Gestalt shift. In one sense, "immediately" is a rhetorical exaggeration, a hyperbole; in another sense, it is not. We may not instantly see the dynamic possibilities of the molecule; but once we do, our perception must be immediate. (1996, 64)

Throughout *The Rhetoric of Science*, I attempt to draw philosophical conclusions from rhetorical analysis.

Campbell, Bazerman, Myers, and I are not the only first-generation rhetoricians to make substantial contributions to the literature. To this list, we must add at least four more: Jean Dietz Moss, Lawrence J. Prelli, and the coauthors Carol Berkenkotter and Thomas N. Huckin. For none of these is rhetoric constitutive of scientific knowledge; Moss is explicit that it is not. She has spent half a scholarly lifetime on placing Galileo's rhetoric firmly within his social and intellectual milieu. In her book, she claims that there was between the *Letter to Christina* and the *Two World Systems* a shift in Galileo's rhetorical strategy. In the former, Galileo is sure that he will find firm proofs of Copernicanism; in the latter, lacking such proofs, he seems to leave the issue open. But Moss shows that Galileo's actual audience was not deceived concerning his real motive in the *Two World Systems*: to propagandize in favor of heliocentricity. Still, for Galileo, rhetoric remains rhetoric; science, science.

Prelli's project rests squarely on the rhetoric tradition: to create a topical rhetoric of science, an inventional system on classical lines that can generate the arguments scientists use. In his analysis of the controversy over whether the great apes can use language, for example, Prelli is concerned exclusively with the type of arguments used by proponents and opponents. In his discussion, he shows that the opponents of research into ape language rely on the scientific norm of universalism as an argumentative strategy; they accuse the ape researchers of systematically avoiding the judgment of peers and interfering subjectively with the outcomes of experiments. Because of these breaches, they affirm, the researchers are not really scientists. The proponents of such research, on the other hand, argue by using a counter-norm, a kind of particularism: their research must use such methods because it is crucial in such experiments to gain the trust of the subjects involved. Theirs is the only way to do simian linguistics. Prelli is not concerned with the substantive issue: is science universalist or particularist? He is concerned only with the rhetorical use of these values to persuade others.

The scholarship of Berkenkotter and Huckin is, in equal measure, social scientific and rhetorical. In their chapter on the news value of scientific articles, for example, they provide us with a genuine insight in their explanatory synthesis of such apparently disparate features of scientific articles: their titles, abstracts, and subheadings, and the structure of their introductory and discussion sections. In the view of these two scholars, the development of this suite of features has the single explanation that each has evolved to increase the news value of the scientific article: "Our analysis of the 350 journal articles in our corpus supports this hypothesis. The 12 scientific journals that we examined have all modified their genre conventions in a way that foregrounds the most important findings of an investigation, that is to say, promotes news value" (1995, 33). For example, they note a significant change in title syntax. A typical title of thirty years ago was: "On the Specificity of DNA Polymerase." As of 1989, however, a title might well encapsulate the claim in a full sentence in the manner of a newspaper headline: "Tumor Necrosis Factor (TNF) Is Induced by *Candida albicans*: The Role of TNF in Fibrinogin Increase" (1995, 33, 54). But for Berkenkotter and Huckin, it is not the science but the way the science is conveyed that is rhetorical.

Rhetoric of Science: Colonization

When I claim that rhetoric colonized other disciplines devoted to science studies, I do not mean that rhetoricians did; I mean, rather that a new climate of opinion emerged, signaled by the linguistic turn in philosophy, exemplified by Ludwig Wittgenstein and John Austin and popularized in 1979 by Richard Rorty's *Philosophy and the Mirror of Nature*. Accordingly, it is no accident that the 1980s, which saw the creation of rhetoric of science, also saw a rhetorical consciousness emerge in disciplines unconnected with the rhetorical tradition. How else can we explain the outburst of rhetorical activity in relation to science by scholars who came from different disciplines and were not, generally, in communication? Participating in this activity were the literary scholar Wilda Anderson; the economist Deirdre (formerly Donald) McCloskey; the anthropologist Emily Martin; the philosophers Marcello Pera, Philip Kitcher, and Ernan McMullin; the linguists John Swales, M. A. K. Halliday, and Ken Hyland; the historians Bruce Hunt and Evelyn Fox Keller; the sociologists Ricca Edmundson and Richard Harvey Brown, and the library scientists Bryce Allen, Jian Qin, and F. W. Lancaster.

Considerations of space limit the number of such figures I can discuss: Allen, Qin, and Lancaster and Swales and Halliday must stand for those whose analyses eschew epistemic implications and Hunt, Brown, and McCloskey for those whose analyses embrace those implications. In "Persuasive Communities: A Longitudinal Analysis of References in the *Philosophical Transactions* of the Royal Society, 1666–1990," Allen, Qin, and Lancaster trace both the pattern and content of the citations in this long-lived international journal. From their abundance of data, they are able plausibly to infer a multiplicity of conclusions. Of persuasive communities in science, they surmise that they shifted from small to large and from personal to impersonal: "Perhaps nowhere is it easier to observe, what persuasive communities are in the seventeenth and eighteenth centuries than in the brevity of many of the bibliographic references from that period. It seems that . . . the authors and works cited were like good friends, easily identifiable by shortened names" (1994, 291).

The authors also infer from citation analysis that the primary medium of scientific communication shifted from books in the seventeenth and eighteenth centuries to journal articles in the nineteenth and to journal articles, conference proceedings, and technical reports in the twentieth. The authors then trace the shift in scientific productivity from Europe to America and track the rise of Soviet science. In addition, they follow the eighteenth-century shift in the language of science from Latin to the various vernaculars. In the latter part of the twentieth century, they note a further shift to English as the international language of science. Finally, and perhaps most important, by treating citations as rhetorical features, the authors are able to "measure" the "rate of change" of persuasive communities; slow in the seventeenth and eighteenth centuries and rapid in the nineteenth and twentieth. In connection with this trend, they venture a cautious prediction concerning the increased tendency toward obsolescence in scientific publications: "If the present trend continues, the median age of the persuasive community may overtake the time required for review and publication of traditional printed communications media. This would lead to increased pressure to adopt speedier means of formal communication in science" (1994, 304).

Swales analyzes the introductions to scientific articles, the classical realm of arrangement or taxis. He discovers a typology of three core steps, whose combined purpose is to create a "research space": (1) establishing an intellectual territory, (2) establishing a niche in that territory, and (3) occupying that niche. In the first step, scientists establish a territory by claiming that the discovery being reported is of interest to the discourse community or by introducing a general technical topic. The next step normally involves acknowledging a contradiction or inconsistency or controversy, identifying a gap in present knowledge, posing a new question or questions, or building on an existing scientific tradition. The third step involves stating the purpose or objectives of the present research, summarizing the principle findings or gist of the solution, providing a road map for the accompanying text, or some combination of all three. Swales notes that not all introductions employ all three moves or present them in the order given. The emergence of the three-step introduction, Swales asserts, is a key element in the shift in scientific articles from an organization emphasizing narrative—what the scientist did—to one emphasizing argument—what the scientist proved.

In "On the Language of Physical Science," Halliday shifts the focus of rhetorical analysis from the organization of scientific articles to their style. He proposes a theory of the development of scientific syntax, the classical realm of lexis. In scientific prose, he notes, nominalization, a normal transformation in language, becomes, virtually, an obsessive concern. In ordinary English, we might say: "Cracks grow at rates that depend on two factors: the chemicals we involve and the degree of stress we apply." In scientific English, we say: "The rate of crack growth depends not only on the chemical environment but also on the magnitude of the applied stress." The two pronouns and three of the four verbs have disappeared.

This "steady drift toward the nominalized region" (1993, 67), Halliday avers, is neither arbitrary nor random; rather, it is a consequence of "creating a new variety of English for a new kind of knowledge" (1993, 81). The step-by-step progress of science—from one experiment to the next, from theory to experiment, and from experiment to theory—demands a syntactical form equal to tracking this progress with perspicacity. In English, the best way to accomplish this task is to construct two noun clauses with a verb in between to tell us how the second follows from the first. Halliday asks:

> Is this form of language more complex? Not necessarily; it depends how we define complexity. If we take lexical density (the number of words per clause), and the structure of nominal elements (nominal groups and nominalizations), it undoubtedly is more complex. On the other hand, if we consider the intricacy of the sentence structure (the number of clauses in the sentence, and their interdependencies), then it will appear as simpler: mainly one-clause sentences; and likewise with the clausal structure—usually only two or three elements in the clause. (1993, 67)

While Allen, Qin, and Lancaster and Swales and Halliday focus on the form of scientific communication, Hunt, Brown, and McCloskey focus on its content. In "Rigorous Discipline: Oliver Heaviside Versus the Mathematicians," historian Hunt extends the scope of rhetoric of science into the realm of mathematical physics. Hunt chronicles a remarkable event, the rejection of one of Heaviside's papers by the *Proceedings* of the Royal Society. As a fellow, Heaviside was by long custom entitled to acceptance of his papers without the formality of refereeing.

In this particular case, however, strong opposition from the mathematicians muscled civility aside. The paper was sent for review to William Burnside, professor of mathematics at the Royal Naval College in Greenwich. "Detailed criticism of results obtained in this way seems out of place," Burnside said in his acid summary. "They may or may not be true, but the way in which they are arrived at makes them absolutely valueless" (qtd. in Hunt 1991, 84).

So devastating a critique can have only three sources: thoroughgoing incompetence on the part of the author, personal vindictiveness on the part of the referee, or profound ideological differences between the two. The actual source was the last. Heaviside was a physicist first. As a physicist, he believed not in the rigor of mathematical proof but in its usefulness for physics; for him, mathematics was an empirical discipline, a tool. Burnside, however, was a representative of a new breed, the English pure mathematicians, trying to elevate the prestige of their discipline through an exclusive emphasis on deductive rigor. This emphasis becomes, in Hunt's telling, not an essential fact of mathematics, pure or otherwise, but a contingent fact of its history. In Hunt's able hands, rigor is transformed from a mathematical given to a rhetorical device, one means of persuading a particularly constituted audience. In his hand, epistemology becomes historically contingent.

In a chapter on the logics of discovery, sociologist Brown extends my analysis of *Narratio Prima*, the first printed embodiment of Copernicanism (see chapter 8). The avowed purpose of Brown's essay is "to define narratives of conversion as a literary/scientific genre" (26). In building his case that "logics of discovery are narratives of conversion generally, and not just for science" (4), Brown first analyzes fictional narratives such as *Moby Dick*. According to Brown, these are stories of conversion designed to draw the reader into the mores of an alien culture, in the case of *Moby Dick*, the culture of whaling. Brown then moves from such novels to ethnographic narratives. He establishes a commonality between fictions such as Moby Dick and ethnographies; in both, the understanding of an alien culture is the goal; in both, the systems of belief of the narrator and reader are altered as a consequence of a double journey—an actual journey into an alien culture and a metaphorical journey into the self.

At this point in his argument, Brown shifts from ethnography to philosophy. In *Discourse on Method*, he contends, René Descartes is taking two journeys, a real European journey and a metaphorical journey into the self. This double journey has a radical impact on Descartes's system of belief. In the last part of his argument, Brown turns from philosophy to science. In *Narratio Prima*, he contends, Rheticus's journey is also a double one: an actual journey to Copernicus and a metaphorical journey to Copernicanism. These journeys have a significance that is epistemological as well as literary:

> As new perspectives become sedimented in discourse, their symbolic, constructed character fades from awareness, and they become more literal and objective. The new stretched genre becomes a thickened old one. A relativistic epistemological stance gives way to a more realistic posture. Rhetorics of conversion, personal testimony, and individual commitment are replaced by their institutional surrogates, the rhetorics of authority and indoctrination. . . . Stretching toward new paradigms or genres, and thickly encoding those already created, are both needed. Without authority and indoctrination we would have no shared language and no intelligible world. Without narratives of conversion we would not be able to invent anything new. (92)

In a groundbreaking work, *The Rhetoric of Economics*, in a chapter entitled "The Rhetoric of Scientism: How John Muth Persuades," economist McCloskey provides an example of her mastery of rhetorical technique. After a plain-English summary of Muth's important article, McCloskey gives us parallel columns: in the left, she reproduces key sentences from his article, all written in the "scientistic" patois of economists; in the right, she turns this patois into plain English. It is a dazzling performance, demonstrating that nothing of substance is lost in the translation. Next, McCloskey infers that, if nothing of substance is lost, "the appeals to the methods of science are mainly matters of style, arising out of a modernist conversation" (1985, 97). Nonetheless, McCloskey avers, Muth needs these scientistic trappings: How else is he going to meet the expectations of economists, who would not be persuaded unless an argument is fitted out with appropriate jargon and decked out with appropriate mathematical formulae? In a finale to this fireworks display of rhetorical proficiency, McCloskey compares argument in economics with argument in three apparently disparate fields. Under their respective disciplinary skins, she finds, arguments in paleontology, mathematics, and literary history parallel those in economics. No interesting epistemological differences exist.

What is McCloskey's view of rhetoric? In the course of her analysis of Muth, McCloskey says: "When honest and well-informed biologists disagree about the strength of a tendency to inherited altruism among human blood relatives or when honest and informed physicists disagree about the significance of Bell's Theorum, they must be using arguments that do not compel assent in the conclusive way required by the modernist method" (1985, 101). Later in the same chapter, she says: "A science is a class of objects and a way of conversing about them, not a way of knowing the truth" (1985, 105). This would seem to indicate that rhetoric is indeed epistemic in the deepest sense; that it is fully constitutive of science. McCloskey's view of science seems virtually identical to the view espoused in my *Rhetoric of Science*.

Science Itself: The Second Generation

In 1997, Dilip Parameshwar Gaonkar's long essay "The Idea of Rhetoric in the Rhetoric of Science" appeared in my and William M. Keith's collection *Rhetorical Hermeneutics*. In this essay, Gaonkar contended that rhetoric was constitutionally unsuited to the analysis of science for three reasons:

1. Rhetoric's essential character, as defined by both Aristotelian and Ciceronian tradition, consists in generating and giving speeches, not in interpreting them—and certainly not interpreting texts such as scientific articles.

2. As traditionally conceived, rhetorical theory requires a strategic model of persuasive speech, one in which an agent—the orator—controls a communicative transaction addressed to an essentially passive audience. Applied to communication within the complex social structure that is science, such a model is implausible.

3. As a consequence of its traditional focus on production, rather than interpretation, rhetorical theory is "thin." The amount of specification necessary for the practice of public speaking embodied in a handbook like Aristotle's *Rhetoric* is far less than that needed for the practice of criticism. Because rhetoric's central terms—*topos*, *pisteis*, and *enthymeme*, for example—elude precise definition, there are few constraints on their employment. Consequently, no interesting conclusions can be drawn from the "fit" between such terms and scientific texts and interpretive consensus as to their rhetorical import will always remain elusive. (1997, 25–85)

Gaonkar's essay represents a reflective moment in which to meditate on the methodological limitations of the first generation of rhetoric of science, limitations that a second generation of scholarship will address. We may divide this latter literature into works like those of Fahnestock and Leah Ceccarelli that assume that rhetoric is, in some sense, constitutive of scientific knowledge and works like Celeste Condit and Gross, Joseph E. Harmon, and Michael Reidy that make no such assumption.

In *Rhetorical Figures in Science*, Fahnestock undermines our comfortable sense that, aside from metaphor, the figures can be safely ignored by rhetorical theorists and critics, that the study of such schemes as *antithesis*, *incrementum*, *gradatio*, *antimetabole*, *ploche*, and *polyptoton* is the preserve only of pedants. She tells us why, although particular figures have been identified in great numbers over the centuries, a definition of figuration has eluded us. Her wide range of examples—from the Bible, from public address, and, most prominently, from science—suggests that the figures exercise their powers regardless of subject matter; the wide range of languages that are the origin of these sources—classical Greek, Latin, French, German, and English—suggest that they exercise their powers regardless of language. Most important, the attention she pays to the visual as well as the verbal strongly suggests that the figures are not linguistic but conceptual in nature and that, as such, they can serve a resource for invention and an index of conceptual change. This book also gives the lie to the pessimistic implication that many have drawn from Gaonkar's assertion that rhetoric is "thin," that is, that the rhetorical tradition lacks the hermeneutic wherewithal adequate to a robust rhetorical criticism.

Fahnestock shows that we have been remiss in not reading Aristotle carefully enough to see that, for him, the figures are also conceptual in nature. We have been so misled by his correct view in book 3 of the *Rhetoric* that their expression is lexical that we have forgotten the passages in book 2 that trace the figures back to their conceptual origins. In chapter 23, for example, Aristotle identifies an enthymematic *topos* derived from opposites; if it is right to do good to one's friends, then it is right to do harm to one's enemies. This is both an *enthymeme and* an *antithesis*. It is this parallelism between the *topoi* and figuration (which, incidentally, opens the door to the conceptual unification of the *Rhetoric*) that Fahnestock wishes to exploit: "Once the verbal form, the figure antithesis, is recognized as the epitome of an underlying topic reasoning, it becomes possible to use the figure itself as a stylistic prompt or frame for invention" (1999, 52).

This point of view makes plausible the claim that, just as metaphor is a vital factor in invention, so also are *antithesis*, *antimetabole*, *gradatio*, and the other members of Fahnestock's family of chosen schemes. This scope includes invention in science. Fahnestock's demonstration of this truth is particularly impressive in the case of Darwin's *Origin of Species*, because this is a book thrice harvested (by historians, philosophers, and rhetoricians) and ready, one would think, only for the gleaners. Fahnestock shows us otherwise; she demonstrates that our obsessive scrutiny of a single figure, metaphor, has blinded us to the pervasiveness of series reasoning in the *Origin*, whose figural realizations are, most commonly, *gradatio*, a series of related items, and *incrementum*, such a series arranged in order of increasing importance. At the center of this great work, Fahnestock reminds us, are the arguments that "establish the presence, or explain the absence of series across three great axes of variability: form, time, and space" (1999, 117).

Consider a problem of spatial distribution that troubled Darwin because it seemed to undermine an implication of his theory: that in regions intermediate between established species, intermediate species must exist. Instead, there exists a sharp line of demarcation. To solve this problem, this absence of *gradatio*, Darwin invented the ecological niche, the device by which species survive by avoiding competition for scarce resources. The suggestion is that in the case of Darwin, "a distinction between the sciences and other fields depends less on the forms of argument use initially than on the degree of accountability demanded eventually" (1999, 43). It is the assertion that such figures as *gradatio* and *incrementum* are a vital part of invention's generative core regardless of subject matter that constitutes Fahnestock's claim to general attention.

In *Shaping Science with Rhetoric*, Ceccarelli is also innovative methodologically; she transforms her three biological case studies—of Theodosius Dobzhansky, Erwin Schrödinger, and Edward Wilson—into an argument concerning the effectiveness of interdisciplinary persuasion in the sciences. In Ceccarelli's view, Dobzhansky convinced geneticists and naturalists that they shared the same object of study; Schrödinger convinced physicists and biologists that their individual perspectives had a contribution to make to the study of the gene; Wilson, on the other hand, failed to persuade either scientists or humanists to embrace his reductionist vision.

In Ceccarelli's view, the two successful scientist-authors convinced through intelligent rhetorical design. Dobzhansky faced the following problem: geneticists and naturalists, his two audiences, saw the world from very different and, so they thought, incompatible perspectives. Naturalists saw populations moving in space; geneticists saw gene frequencies. Dobzhansky had to convince both that gene frequencies and populations were two "sides" of the same Möbius strip. To do so, he used "topographical maps" introduced by Sewell Wright. Dobzhansky adapted these to suit his rhetorical purpose. When naturalists looked at them, they saw population change influenced by gene frequencies; when geneticists looked at them, they saw gene frequencies influencing population change. By such means, "Dobzhansky not only encouraged the recognition that conceptual harmony was possible but also worked to make naturalists think more mechanistically and geneticists to think less mechanistically" (1999). Ceccarelli asserts that Schrödinger faced an analogous problem in his book *What Is Life?*

> In a book that successfully connected physics and biology, the conflicting linguistic practices of these disciplines would have to be negotiated. The solution that Schrödinger stumbled upon was a reversal of linguistic practices. To make biology intelligible to the physicist, Schrödinger had to rid it of its mysterious vitalist presumptions, describing biology in clear mechanistic language. To make physics intelligible to the biologist, he had to describe the objects and concepts of physics in the language of action that was so familiar to the biologist. (1999, 93)

Schrödinger also wielded another tool that language made available—polysemy, the multiple meanings different readers can give the same key phrase. Of a crucial passage in *What Is Life?*, in which he speaks of living matter obeying "other laws of physics" still undiscovered, Ceccarelli says:

> At least five different interpretive communities were addressed with this passage: physicists who were ready to start a quest for exciting new paradoxes in the virgin territory of biology; biologists who wanted to benefit from an alliance with physicists but still retain some level of uniqueness for the living organism; physicists who were unhappy with the current paradoxes in physics and were interested in finding a new all-embracing determinism; biologists who wanted a quick and easy new tool for revealing the secrets of life; and physicists who believed that the current laws of physics would suffice to explain biological processes. (1999, 108)

Polysemy energized five separate groups of scientists, each interpreting "other laws" differently, each pursuing individual research programs that converged in molecular biology.

While Dobzhansky and Schrödinger succeed, Wilson does not. According to Ceccarelli, Wilson fails to persuade because he refuses to use the conciliatory gestures that come naturally to Dobzhansky and Schrödinger. Instead of promoting the language of compromise, Wilson gives us metaphors drawn from conquest. Instead of using polysemy, instead of encouraging readings appropriate to each interest group, he clearly signals his intention to reduce the humanities and the sciences to a single material base that is subject to exceptionless laws. Schrödinger and Dobzhansky are diplomatic; Wilson is uncompromising.

Although *Rhetorical Figures in Science* and *Shaping Science with Rhetoric* implicate rhetoric in the constitution of knowledge, two other books—Condit's *The Meanings of the Gene*, and *Communicating Science* by Gross, Harmon, and Reidy—do not engage epistemic issues. Each is innovative in its combination of rhetorical analysis with methods derived from the social science. Condit's intellectual quarry is the public perception of eugenics and genetics. But a successful hunt requires grounding in sampling and statistical techniques: "the backbone of this study . . . consisted of 653 magazine articles drawn from the *Reader's Guide to Periodical Literature* from 1919 to 1995" (1999, 260). The content of these—and of supplemental sources—was coded and analyzed. In following these procedures, Condit is concerned with matters of statistical significance and interrater reliability. But statistical analysis is not the endpoint of her task. She also asks what her data mean, a task requiring both critical intelligence and rhetorical analysis. She makes her methodological claim explicit: "All critics can be assisted by quantitative methodologies when those methodologies are understood as counting tools, embedded in the critical project, rather than as overarching frameworks that constrain critical thought within a hypothesis-testing method. It is possible to use numbers in a postpositivistic fashion" (1999, 257).

Content analysis allows Condit to make statements like the following: "The movement of genetic information and technologies out of the scientific research laboratory and into general commerce was accompanied by a substantial increase in the quantity of ethical debates about the applications of genetic information. Almost half (45 percent) of the articles in the new era raised ethical concerns of some sort (this was a statistically significant increase from the previous era: see appendix 2, table 5)" (1999, 188). As a consequence of rhetorical analysis, she can also make statements like the following:

> The accommodation of genetic preference and equal humanity was accomplished in large part by moving negative descriptors to the molecular level. Instead of defective people, we now had "healthy genes" and "defective ones" . . . Where an article in the previous era has described a woman with Huntington's chorea as a "culprit" who had passed her genes to many children, articles now referred to "culprit genes" rather than placing the blame on individuals . . . Instead of abnormal persons, we heard about "abnormal genes." (1999, 190)

In *Communicating Science*, Gross, Harmon, and Reidy also combine rhetorical analysis with social science methods. Their goal is sweeping: to provide the reader with a rhetorical history of the scientific article from its seventeenth-century beginnings to the present. Accordingly, their generalizations are the result of "an analysis of 1,804 short passages for style and 430 whole articles for presentation and argument" selected randomly from three languages, English, French, and German (2002, viii). As with Condit's work, these texts are subjected both to statistical and rhetorical analysis. The first procedure yields statements like the following:

> In the 20th century, we find the scientific article growing considerably more uniform across national boundaries and scientific disciplines. Most striking of all, "scientific English" has become the international discourse of science, which involves not only a specific language but also a suite of stylistic features: relatively short, syntactically simple sentences containing complex noun phrases with multiple modification, verbs in the passive voice, noun strings, technical abbreviations, quantitative expressions and equations, and citational traces. French and German scientific articles also conform to this basic style, with some variations: for example, French and German scientific prose has resisted the noun string; each of these languages employs several grammatical structures that avoid the passive voice yet maintain an objective tone. Whatever the language or discipline, the style is streamlined to focus the reader's attention

> on the things of the laboratory and the natural world beyond the printed page, rather than to draw attention to the text itself or its author. (230)

Such generalizations, however, cannot capture the flavor of a changing scientific prose. For that, rhetorical analysis is necessary. For instance, an early botanical article by Martin Lister is analyzed as follows:

> Lister's lack of interest in interpretation, along with his assumption that the workings of nature can be revealed by categorizing and organizing appearances, not only accounts for the virtual lack of substantive arguments in his article, but also leads to an unwavering faith in the scope of his method. In effect, his method constitutes an argument against making arguments. His investigatory presuppositions inform Lister's notion of progress as a gradual and systematic unfolding of the facts of the natural world as they appear to the senses, unrelated to theory. It is for this reason that he can present before the public an article gathered from "*some Papers, which belong'd to a Treatise of Vegetation; they were most of them made about Thirty Years Ago*" . . . In effect, Lister believes that what he had discovered as a young man is not in need of revision, but of expansion. (22)

Communicating Science has two final features of methodological interest: it expands the scope of text to include graphics, the tables, line graphs, photographs, and drawings that are so central to scientific communication, and, in its final chapter, it sketches out an evolutionary theory of the rhetorical development of scientific prose and visuals.

Implicitly or explicitly, these works of the second generation confront and defeat Gaonkar's three theses. While remaining firmly within the rhetorical tradition, they do not suffer from the vagueness of a theoretical terminology tethered to the needs of production, nor do they exalt the individual author as master rhetorical strategist.

2 The Justification of Rhetoric of Science

It may seem mere common sense that classical rhetoric, whose source and first object of study was public utterance in the Greek city-states, is not an appropriate framework within which to view science, quintessentially the product of the modern world in all of its complexity. Can a translation really be possible that will turn classical rhetoric into a system of analysis that responds intelligently to scientific texts without distorting either those texts or classical rhetoric itself beyond recognition? Dilip Parameshwar Gaonkar puts the matter succinctly:

> A putative translator . . . must realize that the translation involves not simply a technical vocabulary but an ideology: a view of speaker as the seat of origin rather than a point of articulation, a view of strategy as identifiable under an intentional description, a view of discourse as constitutive of character and community, a view of audience positioned simultaneously as "spectator" and participant," and finally, a view of "ends" that binds speaker, strategy, discourse, and audience in a web of purposive actions. (qtd. in Gross and Keith 1997, 32–33)

Gaonkar's challenge can be met only if the ideology of classical rhetoric is separated from its technical vocabulary, its kit of critical tools. This separation can be legitimately accomplished, however, only if there is no dependence of that vocabulary on the ideology Gaonkar so tellingly reveals. And there is none. *Stasis*, the three "proofs" of *ethos*, *pathos*, and *logos*, the five canons of invention, and the general and special topics—none of these depend on any aspect of the classical ideology outlined by Gaonkar.

Gaonkar also criticizes the vocabulary itself. While its terms may be perfectly adequate as an aid to the creation of speeches, he avers, they are seriously deficient as part of any system adequate to textual interpretation. Their application simply burdens us with an esoteric vocabulary; we barter insight for pedantry.

The ease with which they apply promiscuously to any text is a danger signal, an index of their essential vapidity as intellectual tools.

But the classical vocabulary of analysis does not require global renovation to create the precision necessary for cogent criticism; indeed, the very plasticity of classical vocabulary permits the local refashioning necessary for a refinement appropriate to scientific texts. A problem that Gaonkar calls "thinness" can be solved by a vigorous dialectic exchange between the terms of the classical vocabulary and the texts in question. The chapters of this book will represent, I hope, instances of the fruitfulness of such an exchange.

The separation of the ideological from the technical allows me to employ the vocabulary of classical rhetoric when my theory of interpretation is not classical, when I feel constrained to use alternate positions as frameworks for analysis: for peer view, Habermas's ideal speech situation; for occupational medicine, Foucault's theory of the relationship between knowledge and power; and for priority disputes, Victor Turner's concept of social drama. Such choices are a consequence of the understandable limitations of classical theory, its relentless focus on the creation of public speech. But this rejection of the classical theory need not be general. The work of René Descartes and of Isaac Newton fits perfectly well within the ideology of classical rhetoric, that is to say, in the relevant dimensions, their work and its dissemination are perfectly analogous to Cicero's. Moreover, those who rely on classical rhetoric need not hark back as far as Cicero; the theory and vocabulary on which he relied have been legitimately extended by Chaim Perelman and Lucie Olbrechts-Tyteca; their *New Rhetoric* is well within bounds. Theirs is a deepening of the tradition; their concepts of audience and presence in particular are useful additions to the classical armamentarium. This is not to say that the science of Descartes and Newton cannot, under the aegis of another theory, be fruitfully treated as points of articulation in the postmodern manner to which Gaonkar alludes.

It is well here to deal with another important criticism of the application of rhetorical analysis to the sciences. Critics have pointed out that the focus on texts, even in their broadest sense—a sense that includes both diary and notebook entries and graphics—constitutes a serious disciplinary limitation, as in the sciences, texts depend heavily on a set of practices well outside the scope of rhetorical analysis. In responding to this allegation, we need first to point out that communication in science *is* a practice on a par with any other. Secondly, we need to acknowledge a legitimate division of intellectual labor. It requires no humility to claim that rhetorical analysis, like sociological, historical, or philosophical analysis, is limited in scope; indeed, it is such limitations that provide the focus that gives disciplines their strength: the claim, after all, is not that rhetoric is co-extensive with science studies but that its application sheds a different and valuable light on the sciences.

It remains in this chapter to show how comfortably the central vocabulary of rhetorical analysis applies to scientific texts and, in so doing, to point out some significant likenesses between the products of classical rhetoric and scientific texts. By way of contrast, the following chapter will deal with the very real differences between genres to which classical rhetoric more easily applies and science, the hardest case. Together, these two chapters should help locate the place of rhetoric in science studies.

Invention: Stasis

At any time, in any science, scientists must make up their minds about what needs to be explained, what constitutes an explanation, and how such an explanation constrains what counts as evidence. When scientists think about matters of explanation, they are deciding what it is to do science. In rhetorical terms, they are using *stasis* theory, an established part of invention, a set of questions by means of which we can orient ourselves in situations that call for a persuasive response. In courtroom arguments, we consider whether an act was committed (*an sit*); whether it was a offense (*quid sit*); whether the offense was justified in some way (*quale sit*). In the analysis of law, these *staseis* have a central role; in the analysis of science, their centrality is equally apparent.

An sit. In the sciences, what entities really exist? Does phlogiston? Do quarks? Before Albert Einstein's papers on Brownian movement, the existence of atoms was in question; after, their existence was regarded as confirmed.

Quid sit. Given that certain entities exist, what is their exact character? From antiquity, light has been steadily the subject of scientific scrutiny. Is light Aristotle's alteration in a medium, Descartes's pressure, Newton's particle, Thomas Young's wave (another alteration in a medium), or the zero-mass particle of quantum electrodynamics?

Quale sit. Even if the character of an entity or phenomenon remains roughly the same, the laws governing it may be radically different: the same law of refraction that is for Newton the result of deterministic forces acting on minute particles is for Richard Feynman the product of probabilistic ones acting on zero-mass particles.

Particular scientific texts emphasize particular staseis. Although Einstein incidentally established the physical existence of atoms, he was mainly concerned with the *quale sit* of Brownian motion. Papers in evolutionary taxonomy establishing a new species mainly support the *quid sit* of existence but are also concerned with the *quale sit* of evolutionary theory. In every case, the staseis focus the scientists' attention on a particular aspect of the problem before them: Newton and Descartes, for instance, were both concerned with the nature of white light, the *quid sit*.

There is a final stasis applicable to both rhetoric and science: what particular body has jurisdiction. Whether something is a scientific theory depends on who is doing the judging. Newton's formulation of the theory of light remained the same throughout his career. At first, it was rejected; its later acceptance depended not on any alteration of the theory but on a change of jurisdiction. In the first court of opinion, Newton was judged by rules of others' making; in the final court, the rules were Newton's own.

Jurisdiction is also important in adjudicating the relationships between science and society. At what point do decisions cease to be internal to a science? The Inquisition saw itself as an appropriate arbiter of all knowledge, including Galileo's scientific theories. In modern times, this determination is usually made by the scientific community, but even contemporary American courts see themselves as proper judges of the social impact of recombinant DNA research.

At any one time, in any one science, there are proper and improper ways to respond to the first three staseis. For Aristotle, for example, the phenomena in need of explanation are those that naturally present themselves: for instance, what accounts for the motion of a stone released from the thrower's hand? This is a case of violent motion, a movement whose efficient cause is the application of a force to an object, overcoming its material cause, its *gravitas*. There is no violent motion without direct contact: a stone thrown in the air continues its motion after it leaves the thrower's hand only because of the impulsive power of the air directly behind it. The stone's initial trajectory is the formal cause of this violent motion. At the height of that trajectory, natural motion takes over; the stone begins to fall, seeking its natural place, the final cause of its motion. The material cause is again the stone's *gravitas*, its formal cause the downward trajectory itself, its efficient cause the distance from its natural place. For Aristotle, scientific explanation is essentially qualitative, according to the four causes; mathematics has no place in physics.

In his *Principia*, Newton escapes the Aristotelian *an sit*; in that work, he no longer takes the traditional topic of motion as the aim of explanation. For Newton, it is not motion but change of motion that requires explanation. Motion itself—intuitively, the natural puzzle—is an aim of explanation no longer in the realm of science. Moreover, Newtonian explanations can be scientific that do not enumerate all four causes. The material cause of change of motion is largely bracketed and its final cause assigned to theology. The efficient and formal causes are privileged, and the formal cause is given a mathematical interpretation: change of motion is explained according to strict mathematical relationships among such nonobservables as force and mass. Such relationships permit quantitative solutions to problems in physics. Wherever possible, these problems have an experimental realization: in principle, Newton will assert only what he can observe under experimental constraints or infer directly from controlled observation.

Because the presuppositions of Aristotle and Newton were opposed and because their notions of evidence and explanation seriously diverged, the sciences they created differed radically. Differently interpreted, the staseis can lead and have led to radically different conceptions of science. Because they precede science, the province of these interpretations cannot be science; even those opposed to a rhetorical approach to science itself must admit that these presuppositions are legitimately the subject of rhetorical analysis.

The Common Topics

The common topics are a staple of classical rhetorical invention: comparison, cause, and definition. These and their fellows are the traditional places where rhetoricians can find arguments on any given topic. These same common topics are also an important source for arguments in science. In his *Opticks*, for example, Newton defines a light ray twice. Early in this work, he provides a definition in terms of the observable: light behaves *as if* it were made up of tiny particles. Later, Newton defines light in terms of a hypothesis concerning the constitution of matter: light *may actually consist* of tiny particles. The difference in these definitions reflects a change in persuasive purpose. By means of the first definition, Newton hopes to persuade the skeptical scientist of the truth of his analysis of light; to agree, this scientist need not subscribe to Newton's speculative atomism. By means of the second, Newton hopes that this same scientist will seriously entertain atomism as a scientific hypothesis.

In Newton's optical works, the common topics are used heuristically as well as persuasively: Newton undermines Descartes's analysis of color by means of the topic of comparison: Newton contrasts Descartes's theory with his own incontrovertible experimental results. Concerned about the material constitution of light, he addresses the topic of cause: the sensation of light, he speculates, is evoked when its tiny particles impinge on the retina. In his presumption of the rectilinear propagation of light, he relies on the topic of authority: everybody since Aristotle has taken this as true.

In each case, we might say that Newton defines scientifically, compares scientifically. But in none of these instances is it possible to define a scientific sense for the common topics, qualitatively distinct from their rhetorical sense: these sources for arguments in science and rhetoric do not differ in kind.

The Special Topics

In addition to the common topics suitable to all argument, there are special topics that provide sources of argument for each of the three genres of speeches: forensic, deliberative, and epideictic. Forensic texts establish past fact; they are so named because their paradigm is the legal brief; their special topics are justice and injustice. Deliberative texts establish future policy; their paradigm is the political speech; their special topics are advantageous and the disadvantageous, the worthy and the unworthy. Epideictic texts celebrate or calumniate events or persons of importance; their paradigms are the funeral oration and the phillipic; their special topics are virtue and vice.

Scientific texts participate in each of these genres. A report is forensic because it reconstructs past science in a way most likely to support its claims; it is deliberative because it intends to direct future research; it is epideictic because it is a celebration of appropriate methods. Analogously, scientific textbooks strive to incorporate all useful past science, to determine directions for future research, and to commend accepted methods. But science also has special topics of its own, unique sources for its arguments. Precise observation and prediction are the special topics of the experimental sciences; mathematization is the special topic of the theoretical sciences. But there is considerable reciprocity. In the experimental sciences, mathematization is also a topic and provides arguments of the highest status; and, in the theoretical sciences, at least by implication, arguments from mathematics are anchored in the special topics of prediction and observation.

Are observation, prediction, and mathematization *topics*? Science is an activity largely devoted to the fit between theories and their brute facts; the better the fit, the better the science. Surely, observation, prediction, and mathematization are not topics but means to that end. In prediction, the confrontation between theory and brute fact is at its most dramatic. Einstein's theory of general relativity forecast the never-before observed bending of light in a gravitational field; Francis Crick's theory of the genetic code predicted that an otherwise plausible variant—the codon UUU—would never occur. Both predictions insisted on the participation of nature; nature, not human beings, would clinch the argument. Einstein's theory was confirmed by the bending of stellar light as measured during a total eclipse; Crick's was disconfirmed by the discovery of a UUU codon. In both cases, it seems, we have left rhetoric behind. We seem to be in direct contact with the brute facts as the criterion for theoretical truth: stellar photographs in the first case, instrument readings in the second.

However, this line of argument fails: in neither case did the brute facts point unequivocally in a particular theoretical direction; in no scientific case do uninterpreted facts—stellar positions, test-tube residues—confirm or disconfirm theories. The brute facts of science are stellar positions or test-tube residues *under a certain description*; and it is these descriptions that constitute meaning in the sciences. That there are brute facts unequivocally supportive of a particular theory, that at some point decisive contact is made between a theory and the naked reality whose working it accurately depicts, is a myth. Observation, prediction, measurement, and their mathematization: these are sources for the arguments in science in the same way—exactly the same way—that the virtuous is the source of arguments for the epideictic orator.

Logic: The Structure of Argument

For Aristotle, scientific deduction differs in kind from its rhetorical counterpart. True, both are conducted according to the "laws" of thought. But rhetorical deduction is inferior for two reasons: it starts with uncertain premises, and it is enthymematic: it generally relies on audience presuppositions to supply missing premises and conclusions. Because conclusions cannot be more certain than their premises and because any argument is deficient in rigor that relies on audience participation for its completion, rhetorical deductions can yield at best only plausible conclusions. Rhetorical induction, reasoning from examples, is equally marked as inferior to its scientific counterpart because of its acknowledged inability to guarantee the certainty of its generalizations: examples illustrate rather than prove.

Aristotle notwithstanding, rhetorical and scientific reasoning differ not in kind but only in degree. No inductions can be justified with rigor: all commit the fallacy of affirming the consequent; consequently, all generalizations of experiments illustrate reasoning by example. Deductive certainty is equally a chimera; it would require the uniform application of laws of thought, true in all possible worlds, the availability of certain premises, and the complete enumeration of deductive chains. But of no rule of logic—not even the "law" of contradiction—can we say it applies in all possible worlds. Moreover, even were such universal rules available, they would operate not on certain premises but on stipulations and inductive generalizations. In addition, all deductive systems are enthymematic: the incompleteness of rhetorical deduction is different only in degree not in kind from the incompleteness of scientific deduction. No deductive logic is a closed system, all of whose premises can be stipulated; every deductive chain consists of a finite number of steps between each of which an infinite number may be intercalated (Davis and Hersh 1986, 57–73). Because the logics of science and rhetoric differ only in degree, both are appropriate objects for rhetorical analysis.

Ethos

Scientists are not persuaded by *logos* alone; science is no exception to the rule that the persuasive effect of authority and the values it embodies weighs heavily. The anti-authoritarian stance, the Galilean myth canonizing deviance, ought not to blind us to the pervasiveness of *ethos* as a source of scientific conviction. Indeed, the progress of science may be viewed as a dialectical contest: on the one hand, we have authority and its values, an authority sedimented in the training of scientists and reinforced by social sanctions; on the other hand, we have the innovative initiatives without which no scientist will be rewarded.

Innovation is the raison d'être of the scientific paper; yet, in no other place are scientific authority and its values more clearly revealed. By invoking the authority of past results, the initial sections of scientific papers argue for the importance and relevance of the current investigation; by invoking the authority of past procedure, these sections establish the scientist's credibility as an investigator. All scientific papers, moreover, are embedded in a network of authority relationships: publication in a respected journal; behind that publication, a series of grants given to scientists connected with a well-respected research institution; within the text, a trail of citations highlighting the paper as the latest result of a vital and ongoing research program. Without this authoritative scaffolding, the innovative core of these papers—their sections on results and their discussions—would be devoid of significance.

At times, the effects of scientific authority can be stultifying: collective intellectual inertia blocks the reception of heliocentric astronomy; Newton's posthumous authority retards the reemergence of the wave theory of light. At other times, times perhaps more frequent, authority and innovation interact beneficially: consider heliocentric astronomy between Nicolaus Copernicus and Johannes Kepler, the theory of light between Descartes and Newton, the concept of evolution in Charles Darwin's early thought; in each of these cases, we see the positive results of the dialectical contest between authority and innovation. These examples alert us that there is no necessary conflict between originality and deference. One of the persuasive messages of authority in science is the need to exceed authority; indeed, the most precious inheritance of science is the means by which its authority may be fruitfully exceeded.

At the root of authority within science is the relationship of master to disciple. To become a scientist is to work under men and women who are already scientists; to become a scientific authority is to submit for an extended period to existing authorities. These embody in their work and thought whatever of past thought and practice is deemed worthwhile; at the same time, they are exemplars of current thought and practice. In their lectures, they say what should be said; in their laboratories they do what should be done; in their papers, they write what should be written.

So long as science is taught as a craft through extended apprenticeship, its routes to knowledge will be influenced by the relationships between masters and disciples. The modern history of heliocentricity is a progress from epicycles to ellipses. But this theoretical development is realized only through a chain of masters and disciples, surrogate parents and adopted children: Copernicus and Rheticus, Maestlin and Kepler. By this means, research traditions are founded, and the methodological and epistemological norms that determine the legitimacy of arguments are passed on as tacit knowledge.

These norms must be integrated into a broader context, one concerned not with facts but values. Modern science is a uniquely European product barely four centuries old, a product whose rise depended on a general transformation. Its wellspring was the widening conviction that the eventualities of the natural order depended not on supernatural intervention but primarily on the operation of fixed laws whose preferred avenue of discovery and verification was quantified sensuous experience. These new preferences established a central Western task: to control nature through an understanding of its laws. To this task, the specific values of science—such as the Mertonian norms of universalism and organized skepticism—are instrumentally subordinate. Equally subordinate are the values on which theory choice depends: simplicity, elegance, and power.

An examination of the forms of authority within science reminds us that the values of science cannot be separated from the social context in which they arise: the early members of the Royal Society decided what science was, how it would be accomplished, how validated, and how rewarded. We need also to be reminded of another set of authority relationships: those between science and society at large. It was the paradoxical promise of early science that it would benefit society best when wholly insulated from larger social concerns. This tenet becomes difficult to justify in an age of nuclear power and gene recombination. Justification is especially difficult when science converts its exceptional prestige into a political tool to protect its special interests, perhaps at the expense of the general interest. The history of occupational medicine is a case in point, the subject of chapter 10.

Pathos

Emotional appeals are clearly present in the social interactions of which science is the product. An examination of these interactions reveals the prominent use of such appeals: the emotions are plainly involved, for instance, in peer-review procedures and in priority disputes. Anger and indignation are harnessed in the interest of a particular claim; they are part of the machinery of persuasion. When science is under attack, in cases of proposed research in controversial areas, emotional appeals also become central. The instance of proposed research in gene recombination is a good example of the fundamental involvement of science in issues of public policy and of the deep commitment of scientists to a particular social ideology.

In addition, the general freedom of scientific prose from emotional appeal must be understood not as neutrality but as a deliberate abstinence: the assertion of a value. The objectivity of scientific prose is a carefully crafted rhetorical invention, a nonrational appeal to the authority of reason: scientific reports are the product of verbal choices designed to capitalize on the attractiveness of an enterprise that embodies a convenient myth, a myth in which, apparently, reason has subjugated the passions. But the disciplined denial of emotion in science is only a tribute to our passionate investment in its methods and goals.

In any case, the denial of emotional appeal is imperfectly reflected in the scientific texts themselves. In scientific papers, the emotions, so prominent in peer-review documents and in priority disputes, though far less prominent, are no less insistently present. In their first paper, James Watson and Crick say of their DNA model that it "has novel features which are of considerable biological interest" (1953, 737). In his paper on the convertibility of mass and energy, Einstein says: "It is not impossible that with bodies whose energy-content is variable to a high degree (e.g., with radium salts) the theory may be successfully put to the test" (1952, 67–71). In these sentences key words and phrases—"novel," "interest," "successfully," "put to the test"—retain their ordinary connotations. Moreover, in Watson and Crick, "considerable" is clearly an understatement: the topic is the discovery of the structure of the molecule that controls the genetic fate of all living organisms.

In short, because science is a fully human enterprise, rhetorical *ethos*, *pathos*, and *logos* are naturally present in scientific texts. Science can constrain but hardly eliminate the full range of persuasive choices on the part of its participants.

The Order of Arguments

In science, the order of arguments is given short shrift. It is hardly noticed; yet, arrangement has always been important. Realizing its powerful effect, Newton cast his physics and recast his optics in Euclidian form. Indeed, during the four centuries of modern science, order has become more rather than less important, more rather than less rigid. Currently, form is so vital a component that no paper can be published that does not adhere closely to formal rules. The order of arguments of scientific papers has become so inflexible that even experienced scientists occasionally chafe under its restrictive principles: results in this section, discussion in that. But when P. D. Medawar, a scientist of wide influence, put his Nobel weight behind a mild reform—putting the discussion section first—his arguments were ignored rather than answered (1964, 42–43).

Yet, nothing is more artificial than the form of scientific papers. Experimental papers, for example, are equally reports and enactments of a set of values inherent in experimental science: the unproblematic progress from laboratory results to natural processes. It is of no consequence that such progress is far from unproblematic or that the philosophical bases of this version of the scientific method have long been undermined. In experimental reports, the order of arguments behaves as a sacred given.

There is another aspect of the order of arguments, one even more central to the operation of science. Aristotle's decision to privilege the proofs of logic and mathematics, to except them from the province of rhetoric, was itself rhetorical, a decision in favor of certain forms of order, a choice that rested on their presumed correspondence to the laws of thought. It is a truism that logical and mathematical proofs are purely matters of syntax, of form, austere tributes to the power of pattern to evoke the impression of inevitability:

All A is B
All C is A
All C is B

Like all syllogisms, *Barbara*, the paradigm syllogism of science, is sound only by virtue of its form. So paradigmatic of absolute conviction have the forms of logic become, so binding has logical necessity seemed, that its force has been attributed to arguments in the natural sciences and, even, in the humanities: we speak of physical necessity and moral necessity as if they and logical necessity were precisely analogous (Perelman and Olbrechts-Tyteca 1971, 193–260). Unlike logical necessity, its physical and moral counterparts are contingent; in all cases, the facts might have been otherwise; likewise, the conclusions derived from these facts.

Style

From the beginning, stylistic choices in modern science have been deliberately trivialized: in the words of Bishop Thomas Sprat, the first historian of modern science, its communications must "return back to the primitive purity and shortness, when men delivered so many things in an equal number of words" (1667, II:xx). In such a program, the schemes and tropes of classical rhetoric are rigorously to be eschewed. Nouns stand for natural kinds; predicates for natural processes. Syntax, the structure of the sentence, is only the reflection of reality, the causal structure of the world.

Scientific style remains oxymoronic at its core: modest in its stylistic resources, heroic in its aim—nothing less than the complete description of reality. Accordingly, tropes such as irony and hyperbole must be barred; they draw attention away from the working of nature. Stylistic devices such as metaphor and analogy cannot be condoned; they undercut a semantics of identity between words and things. Should scientific prose favor the active or the passive voice? This quarrel over schemes—over the appropriate surface subject of scientific sentences—masks essential agreement among the antagonists. Regardless of surface features, at its deepest semantic and syntactic levels, scientific prose requires an agent passive before the only real agent, nature itself. By means of its patterned and principled verbal choices, science begs the ontological question: through style, its prose creates our sense that science is describing a reality independent of its linguistic formulations.

Despite these strictures, tropes such as irony and hyperbole do appear regularly in scientific reports, belying the alleged reportorial nature of these texts and underscoring their persuasive purpose. Although the official view is that metaphor and analogy have only a heuristic function, that they wither to insignificance as theories progress, tropes are central to the scientific enterprise, and some never disappear. In the *Origin of Species*, for example, a central argument is the analogy between artificial breeding and natural selection. This analogy was not abandoned as the theory matured; it was, instead, the means by which the theory has been maintained and extended. Analogy is also central to the whole enterprise of experimental science: laboratory experiments are scientifically credible only if there is a positive analogy between laboratory events and processes in nature.

While the order of scientific arguments has an epistemological role, designed to show us how science works; the style of scientific arguments has an ontological role: the syntax and semantics of scientific utterances are designed to reveal the causal structure of the world.

Despite their very real similarities, the differences between the texts of the sciences and those to which classical rhetoric easily applies are real and important. These differences are not revealed, however, by the varying susceptibility of these texts to rhetorical analysis; rather, they are revealed precisely as a consequence of their joint susceptibility. In the next chapter, I will explore these differences by focusing on one specific difference: the use of analogy in oratory, philosophy, and science. I think that this comparison will dramatize interesting differences among three very different ways of knowing and understanding the world. If successful, these two chapters will have made a case for the rhetorical analysis of the sciences while insisting that the sciences share rhetorical characteristics among themselves that make them unique.

3 The Kind of Rhetoric Science Is

Analogy is traditionally a device central to rhetorical proof; its continuing importance in modern oratory is illustrated in Winston Churchill's dictum that "apt analogies are among the most formidable weapons of the rhetorician" (qtd. in Montalbo 1978, 6). As Chaim Perelman and Lucie Olbrechts-Tyteca's long section on analogy reminds us, this method of argument is also pervasive in philosophy and in the human sciences (1971, 371–459). Finally, analogy has had a long history in the sciences themselves. An important device in Aristotle's scientific writing, it is still very much in use today—to call the configuration of the components a genetic code is to make a scientific analogy. I hope to show that each domain—oratory, philosophy, and science—uses analogy differently and that these differences reflect distinctions in the kinds of agreement sought and achieved.

Analogy in Oratory

We can best analyze the use of analogy in oratory by concentrating on the work of an acknowledged master, Franklin Delano Roosevelt. On March 4, 1933, FDR delivered his first inaugural address to a nation in the grip of the Great Depression. A single analogy governs the speech: the relationship between politics and war. Roosevelt introduces this military analogy when he speaks of "terror which paralyzes needed efforts to convert retreat into advance." Later, he refers to the task ahead as "the emergency of a war." The American people, he says, "must move as a trained and loyal army willing to sacrifice for the good of a common discipline," and he asks for "a unity of duty hitherto evoked only in time of armed strife."

> I am prepared under my constitutional duty to recommend the measures that a stricken nation in the midst of a stricken world may require.

> These measures, or such other measures as the Congress may build out of its experience and wisdom, I shall seek, within my constitutional authority, to bring to speedy adoption.
>
> But in the event that the Congress shall fail to take one of these two courses, and in the event that the national emergency is still critical, I shall not evade the clear course of duty that will then confront me.
>
> I shall ask the Congress for the one remaining instrument to meet the crisis—broad executive power to wage war against the emergency, as great as the power that would be given to me if we were in fact invaded by a foreign foe. (1963, 198)

In the first inaugural, the appeal to reason is firmly minimized, as Eleanor Roosevelt discovered to her dismay. She found the inaugural terrifying "because when Franklin got to that part of his speech when he said it might become necessary for him to assume powers ordinarily granted to a President in war time, he received his biggest demonstration." That Roosevelt's audience clearly preferred his strong leadership is indicated by the nearly half a million letters the White House received immediately following the speech (Schlesinger 1959, 1).

The minimization of rational appeal was confirmed at Roosevelt's first news conference four days later. FDR showed that, while his bid for strong leadership was clear, his programs were vague indeed:

> QUESTION: In your inaugural address in which you only touched upon things, you said that you are for sound and adequate [currency]—
>
> THE PRESIDENT: I put it the other way around. I said "adequate but sound."
>
> QUESTION: Now that you have more time, can you define what that is?
>
> THE PRESIDENT: No. (*Laughter*) In other words—and I should call this "off the record" information—you cannot define the thing too closely one way or the other.
>
> (1972, 1:11)

As one Congressman complained, "He is for sound currency, but lots of it" (qtd. in Leuchtenburg 1973, 42).

This attitude toward intellectual analysis runs throughout the speech. Roosevelt's goal is not rational assent based on "a genuine understanding of the position assented to" (Johnstone 1963, 92) but emotional commitment. The listener is asked to re-conceive social reality, to see civilian life in terms of military life, and thus unequivocally to prefer FDR's authoritarian stance. This is no isolated instance of the political uses of this type of analogy. In the intervening years have been other "wars": on cancer, on poverty, on terrorism. In the latter case, at least, the parallel with the Rooseveltian appeal is most striking and—in a word—terrifying. This is because in both cases it provides "good reasons" to extend the bounds of executive power.

Why is this analogy persuasive? Because it is apt, that is, it is made as a consequence of Roosevelt's intuitive grasp of its American significance (Oakeshott 1962, 235; Perelman and Olbrechts-Tyteca 1971, 391). In this analogy, understanding the effectiveness of the military comparison means perceiving the natural fear of invasion of a people only the oldest of whom could remember battles fought within their own borders. To form a bridge between the Great Depression and foreign invasion, the analogy assumes the knowledge and acceptance of Roosevelt's dual constitutional role as president and commander-in-chief. Effectiveness depends also on fear of another sort of invasion, the terror of infection invoked by the adjective "stricken."

Though rational appeal has been minimalized, it would be wrong to call the American response irrational. Because an authentic emotional response "can never be deliberate and carefully considered," we tend to regard our emotions as "irrational and disruptive." Nevertheless, we may be truer to our experience if we see them as judgments of a special sort whose general purpose is to maximize self-esteem. Our general perception of the disruptiveness of strong emotional responses remains accurate; but this perception now arises entirely from the extraordinary exigencies that are the proximate cause of strong emotions, exigencies that exceed the carrying capacity of our usual repertoire of responses. This view makes the emotions purposive, an aspect of our ethical behavior, and an arena that licenses learning from a new point of view (Solomon 1980, 270; 251–52).

This analysis explains the American response to Roosevelt's first inaugural. The required response was assent, an emotional reaction to crisis. While this reaction enhanced self-esteem by indicating that Americans had the means to cope with an unprecedented cataclysm, it did not exclude the acceptance of full responsibility for the consequences of that acceptance: the assent to new social policies that flowed from enhanced political power. Analogy mobilized strong emotions in the interest of the transformation of social reality, a reality that was a precursor to a new set of beliefs whose value as a survival method was viewed as significantly higher than the old set. As a result of this transformation, the military fused with the civilian, which created popular support of the strong executive measures that characterized the federal activity of Roosevelt's first hundred days.

Analogy in Philosophical Argument

For a good portion of their professional lives, Sir Karl Popper and Thomas S. Kuhn argued with each other concerning the meaning of continuity and progress in the sciences. Kuhn believed that scientific revolutions, such as the Copernican, resulted in major changes in the way scientists perceive the world and formulate problems; thus, scientific growth was discontinuous, not progressive, although "at least part of the [preceding] achievement always proves to be permanent" (1970, 25). Popper, on the other hand, believed that "in science (and only in science) can we say that we have made genuine progress: that we know more than we did before" (1970, 57).

In his essay "Normal Science and Its Dangers," Popper uses analogy to assert the possibility of significant communication between rival major scientific systems ("frameworks")—such as the Ptolemaic and the Copernican. Without this possibility, he believes, there can be no scientific progress: "It is just a dogma—a dangerous dogma—that the different frameworks are like mutually untranslatable languages. The fact is that even totally different languages (like English and Hopi, or Chinese) are not untranslatable, and that there are many Hopis or Chinese who have learned to master English very well" (1970, 56).

In his response to Popper, Kuhn accepts the latter's analogy but draws a crucial distinction between the ability to learn another language and the ability to translate from one language to another. This method of refutation is only effective insofar as "it emphasizes the fragility and arbitrary character of the analogy" (Perelman and Olbrechts-Tyteca 1971, 387). We shall see how Kuhn manages:

> I accept the utility, indeed the importance, of the linguistic parallel, and shall therefore dwell for a bit upon it. Presumably Sir Karl accepts it too since he uses it. If he does, the dogma to which he objects is not that frameworks are like languages but that languages are untranslatable. But no one ever believed they were! What people have believed, and what makes the parallel important, is that the difficulties of learning a second language are different from and far less problematic than the difficulties of translation. Though one must know two languages in order to translate at all, and although translation can then always be managed up to a point, it can present grave difficulties to even the most adept bilingual. He must find the best available compromises between incompatible objectives. Nuances must be preserved but not at the price of sentences so long that communication breaks down. Literalness is desirable but not if it demands introducing too many foreign words which must be separately discussed in a glossary or appendix. People deeply committed both to accuracy and to felicity of expression find translation painful, and some cannot do it at all. (1970, 267)

Like languages, rival scientific theories are untranslatable because they "cut up the world in different ways." Just as different languages divide the color spectrum differently—the Hindus, for instance, have one word for both orange and red—so in a science like chemistry, key terms change their meanings after scientific revolutions: "Alloys were compounds before Dalton, mixtures after" (Lakatos 1970, 269).

In this exchange between Popper and Kuhn, analogy is both the agent of intellectual exploration and a vehicle for proof. As a heuristic, it is active, helping to create hypotheses. Nevertheless, analogy also serves a probative purpose: the translation analogy of Popper and Kuhn is as much a part of their argument as Roosevelt's military analogy is part of his. However, in Roosevelt's case, the chief criterion for success is assent regardless of understanding, assent procured mainly through the emotional force of the military comparison. Popper and Kuhn's analogy, on the other hand, is successful only insofar as its force is rational: the assent it legitimates fully depends on the understanding it creates. This difference in aim is fostered by the emotional neutrality of the translation analogy, so different in nature from the emotionally charged comparison used by Roosevelt.

Not that philosophical discourse is free from emotional heat. Popper's prose contains the admonitory locution "a dogma—a dangerous dogma." And Kuhn expresses exasperation: "But no one ever believed they were!" Neither of these locutions, however, is central to the argument. But if the emotions are peripheral to their argument, the personalities of the individual scholars are not. The premium placed on individual intellectual brilliance and the relative absence of unanimity and progress in the humanities and in the human sciences are not unconnected. Physics excludes neither individual brilliance nor acrimonious dissent, but physicists may infer what humanists may not: that the problems on which they are concentrating are connected in such a way that their joint solution will further knowledge.

In philosophy, this feeling is an exception. Popper and Kuhn raised to new heights the level of discourse in their subject. Nevertheless, the unity of a discipline and its progress depend not on the depth of individual intellectual achievement but on significant agreement among its best practitioners; its Poppers and its Kuhns must agree concerning the criteria by which significant theoretical differences can be resolved. And one senses that, in all likelihood, Popper and Kuhn and their intellectual heirs will continue to disagree, that no evidence on the one side will be judged conclusive by the other. As a result, intellectual performance in philosophy will be spurred on by the need to refute the opponents' arguments or to support one's own, to gain victories, recruit disciples, or effect conversions.

Kuhn's rebuttal of Popper within the context of Popper's own analogy illuminates the limits of possible agreement in philosophical argument. Such arguments are not open to everyone. They can occur only within a small group with specialized prerequisites, men and women trained in the complex etiquette of a discipline, comfortable with its styles of reasoning, and able to solve its current problems in a manner judged worthy by its intellectual guardians. In this sense, Popper and Kuhn may be said to agree, for if assent to or dissent from their positions was not based on "genuine understanding" of these positions, their combats would be devoid of intellectual significance (Johnstone 1963, 92).

Nevertheless, this commitment to scrutinize analogical arguments defines the limits of significant agreement in fields that are scholarly rather than scientific. In science, though not in scholarship, this commitment is supplemented by a set of agreed-upon quantitative methods, methods that in the aggregate are capable of settling theoretical disputes. As a result, in science, in contrast to the humanities and human sciences, a reliable level of significant agreement is possible.

Analogy in Science

Despite his best efforts, the seventeenth-century Swiss mathematician Jacques Bernoulli was unable to find the sum of the infinite series of the squares:

$$1 + 1/4 + 1/9 + 1/16 + 1/25 + 1/36 + 1/49 + \ldots = ?$$

Because the numbers in this series become progressively smaller, an approximate solution is readily available. In the following century, another mathematician, Leonhard Euler, found an answer to seven places. Dissatisfied with this approximation, Euler, also a Swiss, made a daring leap by means of analogy: he used algebra designed for finite mathematics to solve an unalgebraic equation, "applying the rules made for the finite to the infinite." This solution by analogy and its accompanying answer of $\pi^2/6$ was in no way a proof. Finally, after much effort, Euler "succeeded in verifying . . . exactly, the value of $\pi^2/6$ for Jacques Bernoulli's series [by a proof that was] hidden and ingenious" (Polya 1954, 21). As this example demonstrates, analogy has a role in mathematics: not probative but heuristic.

The "genetic code" is an example of the heuristic role of analogy in the sciences. In the second quarter of the twentieth century, scientists repeatedly suggested a strict relationship between the inheritable characteristics of plants and animals and the exact structure and sequencing within their chromosomes. The most famous statement of this relationship was an analogy. Erwin Schrödinger, Nobel laureate in physics, in his speculative lecture series published in 1944 as *What is Life?* makes the following assertion: "With the molecular picture of the gene it is no longer inconceivable that [a] miniature code should precisely correspond with a highly complicated and specified plan of development and should somehow contain the means to put it in operation" (1967, 66). In other words, there is an analogy: just as a code conveys information from one human being to another, a genetic code transfers genetic information from the hereditary substance to the protein that forms living matter.

Because of the virtual absence of observational and experimental evidence, neither the alleged code nor its transmission mechanism could be elucidated, and statements like that of Schrödinger must be classed as speculation, not science. Not until 1953 did Schrödinger's genetic-code analogy definitely prove its usefulness. In that year, James D. Watson and Francis H. C. Crick made their Nobel-prize–winning discovery of the structure of the hereditary substance, now identified as deoxyribonucleic acid (DNA), "two helical chains each coiled round the same axis," each chain joined to its partner by a series of base pairs. "If an adenine [A] forms one member of the pair, on either chain, then on these assumptions, the other member must be thymine [T]; similarly for guanine [G] and cytosine [C]. The sequence of bases on a single chain does not appear to be restricted in any way. However, if only specific pairs of bases can be formed, it follows that if the sequence of bases on one chain is given, then the sequence on the other chain is automatically determined." Watson and Crick conclude "that the specific pairing we have postulated immediately suggests a possible copying mechanism for the genetic material" (1953, 738). The nature of this copying mechanism remained to be elucidated.

It soon became apparent that ribonucleic acid (RNA), closely related to DNA, was central to the elucidation of the genetic code. In RNA, uracil [U], not thymine [T], bonds with adenine [A] and is the fourth base, the fourth "letter." Within the decoding process, DNA transcribes the hereditary message onto RNA. By a complex series of mediations, RNA translates the message onto appropriate protein, a message transcribed by units called codons.

Each codon or "word" is three bases long; each has to be "read" correctly in order for the hereditary mechanism to function properly. Crick said a correct reading would be ensured if the genetic code was "comma-free." A comma-free code is so constructed that a correct reading is independent of starting place. Here is a lexicon of twenty codons for a comma-free triplet code (the slashes are for expository purposes only):

> ACA/ACC/ACG/ACU/AGA/AGG/AGU/UCA/UCC/UCG/
> UCU/AUA/AUU/UGA/UGG/UGU/GCA/GCC/GCG/GCU
>
> (Woese 1967, 23–29)

To demonstrate the comma-free nature of this code, let us take the following imaginary short chain:

A	U	U	U	G	A	A	U	U	U	C	A
1	2	3	4	5	6	7	8	9	10	11	12

Given the above lexicon, there are only four codons in this chain: Only the first, fourth, seventh, and tenth positions represent legitimate starts. A start at the second position, for instance, gives the nonsense reading UUU.

Crick and his associates claimed that the genetic code was comma-free. But, despite its theoretical elegance and its ingenious interpretation of the coding analogy, their comma-free code did not stand up to mounting counterevidence. The final blow was rendered by Marshall W. Nirenberg—later a Nobel laureate—and his coworker J. Heinrich Matthaei. They clearly established that the "addition of . . . polyuridylic acid [to a stable cell-free system] specifically resulted in the incorporation of L-phenylalanine into a protein resembling poly-L-phenylalanine" (1961, 1601). Thus, "at least one codon all of whose bases were identical was assigned to an amino acid" (Woese 1967, 27)—the "illegitimate" codon (according to comma-free rules), UUU. This meant that the correct reading of the genetic substance was accomplished not by a comma-free code but by accurate start and stop signals.

This example underlines the heuristic as distinct from the probative use of analogy in the sciences. The analogy that led Crick to the correct interpretation of the structure of DNA also led him to the incorrect formulation of the way the genetic code was read. However, Nirenberg and Matthaei, working with the same analogy, discovered the correct formulation.

Scientific reports and scholarly arguments are alike in the value they place on the heuristic function of analogy and on the rules of inference and evidence with which analogies and the hypotheses they generate must be examined. However, scientific reports have recourse to one additional tool: a complex of quantitative methodologies shared by scientists and central to their verification procedures. For instance, Nirenberg and Matthaei's techniques of separation (ultra centrifugation and electrophoresis) and of tracing (paper chromatography and radioactive isotopes) are well-founded practices the results of which can be questioned only by questioning the conclusions of decades of work by thousands of scientists in dozens of fields. As philosopher Max Black says: "In appraising [scientific] models as good or bad, we need not rely on the sheerly pragmatic test of fruitfulness in discovery; we can, in principle at least, determine the 'goodness' of their 'fit'" (1962, 238).

Of course, the leap from quantitative readings to valid scientific inferences is not automatic. Because of the problematic nature of this link, Nirenberg and Matthaei proceed with caution. They show that their unknown substance shares four key characteristics with "authentic poly-L-phenylalanine" and conclude merely "that polyuridylic acid contains information for the synthesis of a protein having many of the characteristics of poly-L-phenylalanine." Still, the example of the discovery of the genetic code demonstrates the compelling argumentative force of such methods of verification.

At a Biochemical Congress in Moscow, Crick heard about Nirenberg and Matthaei's devastating critique of his comma-free system:

> The audience of Symposium I was startled by the announcement of Nirenberg that he and Matthaei had produced polyphenylalanine (that is, a polypeptide all the residues of which are phenylalanine) by adding polyuridylic acid (that is, an RNA the bases of which are all uracil) to a cell-free system which can synthesize protein. This implies that a sequence of uracils codes for phenylalanine. (Crick, Griffith, and Orgel 1961, 1232)

This discovery forced Crick to reject his comma-free formulation in favor of the alternate hypothesis that "the correct choice [is] made by starting at a fixed point along the sequence of bases three (or four, or whatever) at a time." Such ready agreement demonstrates less fellow feeling than necessary assent to verification based a privileged class of observation statements generated by agreed-upon practices.

Is this effect so central to the sciences susceptible to rhetorical interpretation? Perelman and Olbrechts-Tyteca demonstrate that it is. In their discussion of the dissociation of concepts, they speak of "the 'appearance-reality' pair": appearance or, more generally, term I; reality or, more generally, term II. To Perelman and Olbrechts-Tyteca, "term I corresponds to the apparent," while "term II provides a criterion . . . a rule that makes it possible to classify the multiple aspects of term I in a hierarchy. It enables those that do not correspond to the rule which *reality* provides to be termed . . . erroneous" (1971, 416–17; emphasis in original). If dissociation as defined by these authors is applied to this case, term I is the appearances common to Crick and to Nirenberg and Matthaei, appearances that led to diverse and contradictory speculation; the reality is the product of a privileged class of observation statements derived from agreed-upon practices pointing to a single explanation. Indeed, Perelman and Olbrechts-Tyteca's insight into the characteristics of term II applies directly to my account of the discovery of the genetic code: "Term II profits from its oneness, from its coherence, when set against the multiplicity and incompatibility of the aspects of term I, some of which will be disqualified and marked ultimately for disappearance" (1971, 416–17). The persuasive effect of science becomes just its ability to move from term I to term II *as if* moving from appearance to reality. This is a move unique to the sciences.

A View of Truth Compatible with This Analysis

Oratory, philosophical argument, scientific reports: in each, analogy functions differently. In oratory, the route to assent is primarily through emotional commitment. In philosophical argument and in scientific reports, rational commitment is the primary route; but commitment in science reaches, seemingly, beyond argument. In a book whose topic is the rhetoric of science, we need an account of truth compatible with these very different effects, especially with the firm sense of certainty that is the gift of the sciences. In short, we need a rhetorical account of truth. Such an account need not turn its back on the analytic tradition in philosophy; indeed, the account I offer will bear a strong family resemblance to the views of Willard Van Orman Quine, Donald Davidson, and Hilary Putnam.

"*What is Truth*? said jesting Pilate; and would not stay for an answer" (Bacon 1937, 4). Pilate's question is not skepticism but cynicism, not the doubt that precedes inquiry but a dogmatism that dismisses inquiry as futile. The real question, the skeptical question, has two traditional answers: truth is the correspondence between the sentences we form and states of affairs in the world; alternately, truth is the coherence among sets of sentences.

In a correspondence theory, true sentences correspond to the way the world is: they fit into the world as one piece of a jigsaw puzzle fits into another: "Snow is white" because snow is white. The trouble with this theory is that sentences and the world, unlike jigsaw pieces, are so completely different from each other that it is hard—I would say, impossible—to make real sense of their fitting together. Sentences consist of meaningful sequences of words, while the world consists of objects and events that do not mean but just are: objects and events are capable of meaning only because and insofar as sentences refer to them and attribute meaning to them. Taken by itself, then, a correspondence theory of truth seems to confuse two incompatible orders of being.

In a coherence theory, truth is a relation among sentences that hang (or do not hang) together. To see how this theory works, let us return to the jigsaw analogy, imagining a box that contains all of the pieces of a puzzle that, unknown to us, form an Alpine landscape when fitted together. Unfortunately, this is the same box in which a demonic six-year-old has placed numerous pieces from other jigsaw puzzles. In this analogy, just as we discover the Alpine landscape by finding and fitting together all and only those pieces that form a coherent picture, we discover the truth by finding and fitting together all and only those sentences that form a coherent whole. On its face, however, the coherence theory of truth seems empty: surely, one might say, sentences are true only by virtue of referring to and attributing correct meanings to the objects and events of the world.

Taken by itself, then, each theory of truth is inadequate; taken together, however, they fuse in a definition that can be defended. This is the route taken by Quine, Davidson, and Putnam, and it is the route I will take as well. In his classic essay "Two Dogmas of Empiricism," Quine makes a point central to any view of scientific truth that might be compatible with rhetorical analysis: "Total science, mathematical and natural and human, is . . . underdetermined by experience. The edge of the system must be kept squared with experience; the rest, with all its elaborate myths or fictions, has as its objective the simplicity of laws" (1961, 45). To say that our theories must conform to our experiences is not to say what is in *here* must conform to what is out *there*. It is to say rather that our theoretical sentences—myths and fictions in Quine's provocative formulation—must cohere with our observation sentences, those "on which all speakers of the language give the same verdict when given the same concurrent stimulation" (1969, 86–87). "The thermometer reads 98.6°" might be such a sentence.

This relationship between theoretical and observation sentences is the truth behind the correspondence theory of truth. If the network of scientific texts contains the appropriate theoretical and observation sentences in the appropriate configuration, then this network represents the truth of science, *so far as we know*. This is the truth behind the coherence theory of truth.

Let us look more closely at the clause *so far as we know*. To say that theories are underdetermined by experience is merely to say science never can have enough evidence to demonstrate that a theory is true. At any time, then, additional experiences may undermine existing theories by forming a new configuration of observation sentences supportive of a new theory. Quine dramatizes this ever-present possibility of theory revision, otherwise known as scientific progress, by calling theories as myth or fictions.

Quine's view is largely shared by Davidson and Putnam. Davidson says that "the crucial point on which I am with Quine might be put: all the evidence for or against a theory of truth . . . comes in the form of facts about what events or situations in the world cause, or would cause, speakers to assent to, or dissent from, each sentence in the speaker's repertoire" (1984, 230). This echoes Quine's point about observation sentences. To Putnam, "'Truth' [is] . . . some sort of ideal coherence of our beliefs with each other and with our experiences *as those experiences are themselves represented in our belief system*" (1981, 49–50; emphasis in original). This echoes Quine's point about the coherence among our sentences.

Putnam also says, "*[M]eanings just aren't in the head*" (1981, 19; emphasis in original). This statement is ambiguous. Putnam can mean just what Quine and Davidson mean: that observation sentences mediate between mind and world. But if he means what he says literally, then he is mistaken. When experiences are represented in our belief system, they *are* in our heads and nowhere else. Surely, it is only in our heads that such representations are possible, and it is only on account of such representations that our experiences can mean. While our sentences about the world are caused by objects and events in the world, it is we and not the world who attribute meaning to those objects and events.

Those who resist the notion that science is receptive to rhetorical analysis point to the "brute facts": planes fly, men can't have babies, Snel's Law will not be repealed. Surely, we can't be skeptical of such facts. But a rhetoric of science denies none of these. No theory of physics can ignore flight; no theory of biology can turn its back on sex; no optics can dismiss refraction. Rather, the claim of rhetorical analysis is that the phrase *brute facts* is an oxymoron. Facts are not in the world but in our heads; they are by nature linguistic—no language, no facts. By definition, a mind-independent reality has no semantic component. It can neither mean nor be incorporated directly into knowledge. Incorporation by reference is the only possibility: sentences that are candidates for knowledge must refer to a mind-independent reality in fact or in principle, a reference earned in a manner approved by relevant epistemic communities, in this case communities of scientists who superintend their disciplinary domains. Only such sentences can become part of bodies of knowledge regarded as objective.

Rhetoric of science does not deny that the "brute facts" are critical to science. Low-level generalizations such as Snel's Law, essentially descriptive enterprises such as taxonomy and observational astronomy, are part of every science precisely because no science can afford to ignore such generalizations and such enterprises. Because the "brute facts" are constraints on any theory, they tend to persist, even through scientific revolutions. Whatever the interpenetration of scientific theory and fact, there is a sense in which Ptolemy, Copernicus, Kepler, and Newton explain the same night sky, described in essentially the same observation sentences, sentences such as "On day *y*, planet *x* exhibits a retrograde motion of *z* degrees of arc." What is stable in science is the much-denigrated world of appearances, embodied in observation sentences, the only world with which science must square itself. It is this world with which the edges of the system that is science must be squared.

Putnam is at pains to dismiss the views of those for whom "'true' is just a name for what a bunch of people can agree on" (1987, 17–18). But to say that scientific knowledge represents a consensus concerning the coherence and empirical adequacy of scientific utterances, that the various methods of science are essentially consensus-producing, is not to denigrate science; it is rather to pay tribute to the supreme human achievement that genuine consensus on complex issues—consensus based on mutual understanding—represents, an achievement to which this book, I hope, will bear witness.

From the point of view of rhetoric, the truths of science are not beyond argument; rather, they are achievements of argument; science rests on facts and theories that have been argued into place. To this, Putnam nearly agrees:

> [A]n argument between an intelligent liberal and an intelligent Marxist will have the same character as a philosophical dispute at the end. And we all do have views in religion, or politics, or philosophy, and we all argue them and criticize the arguments of others. Indeed, even in 'science,' outside the exact sciences, we have arguments in history, in sociology and in clinical psychology, of exactly this character. (1981, 112)

But why exclude the exact sciences? Because, unlike their "inexact" counterparts, the exact sciences can "appeal to *public* norms" of rationality, norms that are embodied, presumably, in the various methods of science (Putnam 1981, 111; emphasis in original). But this says only that the exact sciences are the product of a second-order agreement: an agreement concerning what counts as agreement. This guarantees only the firm sense of certainty with which the sciences have made us familiar; it cannot purchase genuine epistemological superiority. By calling some sciences exact, Putnam smuggles into his own rhetoric an epistemological superiority he has not earned through argument.

When scientific truth is a consensus concerning the coherence of a range of sentences, rather than the fit between the facts and reality, conceptual change need no longer be justified on the basis of its closer approximation to that reality. It is instead the natural result of the persuasive process that is science, a persistent effort to renew consensus, despite a constant influx of potentially disruptive sentences. Because this influx often threatens an existing consensus, one in which men and women have a serious professional investment, any new agreement is an achievement. At the limit of professional investment, we have the cluster of theories absolutely central to research programs; in this case, a new consensus is understandably infrequent.

This view of science bears a close resemblance to that in Nelson Goodman's *Ways of Worldmaking* (1978, x). To Goodman, science is less a matter of truth than of making worlds. In the absence of "a ready-made world," waiting to be discovered (1978, 94), new worlds are necessarily constructed from the old: Making is always a remaking that depends heavily on methods sanctioned by tradition:

> Without the organization, the selection of relevant kinds, effected by evolving tradition, there is no rightness or wrongness of categorization, no validity or invalidity of inductive inference, no fair or unfair sampling, and no uniformity or disparity among samples. Thus justifying . . . tests for rightness [a more general term that includes truth] may consist primarily in showing not that they are reliable but that they are authoritative. (1978, 138–39)

Goodman shows us how we can stay within the analytical tradition without privileging either truth or scientific knowledge and without accepting realism in any of its more naive forms. Goodman's view of science as worldmaking also provides us with a philosophy fully compatible with the rhetorical analysis of science.

Let me provide an example of worldmaking. Just a few months after his 1835 visit to the Galápagos, Darwin wrote to his friend the naturalist John Henslow and sees nothing of evolutionary significance: "I last wrote to you from Lima, since which time I have done disgracefully little Nat: History; or rather I should say since the Galapagos Islands, where I worked hard.—Among other things, I collected every plant . . . I paid also much attention to the Birds, which I suspect are very curious" (1985, 1:485).

Three years later, in 1839, already thinking about evolutionary theory, as we know from the *Notebooks*, Darwin mentions a peculiar characteristic of one species of bird—the finches. In his popular travel book, *The Voyage of the Beagle*, he says: "The beak of the Cactornis is somewhat like that of a starling; and that of the fourth subgroup, Camarhynchua, is slightly parrot-shaped. Seeing this diversity and gradation of structure in one small, intimately related group of birds, one might really fancy that from an original paucity of birds in this archipelago, one species had been taken and modified for different ends" (1962, 381). These are hints and guesses; only in retrospect is their world-changing significance plain.

On the Origin of Species was published two decades later. From that time on, we had in our possession a plausible theory to account for these differences, just as, after the more recent fieldwork of Peter Grant and Rosemary Grant, we had actual evidence of finches evolving, the distribution of beak size varying, as it were, before our eyes. As a consequence of this theory and its accompanying abundance of evidence, to visit the Galápagos today is to view at close hand an isolated and protected area in which numerous species—finches, sea lions, iguanas—coevolve in response to the selection pressures of a changing climate and changing interrelationships.

To visit the Galápagos today is, equally, to return to the archipelago that Darwin visited and to enter the new world that Darwin was instrumental in making. Inevitably, we view the islands through the lens of evolutionary theory, a *coherent* set of sentences that, we believe, *corresponds* exactly to the facts on the ground.

Part Two

The Limits of the Rhetorical Analysis of Science

4 The Possibility of a Complete Rhetorical Description of Science

Taxonomy as Example

In his famous and influential paper central to twentieth-century debates about quantum reality, written with Boris Podolsky and Nathan Rosen, Albert Einstein insists that the following criteria must be satisfied by any physical theory pretending to the truth:

> Any serious consideration of a physical theory must take into account the distinction between the objective reality, which is independent of any theory, and the physical concepts with which the theory operates. These concepts are intended to correspond with the objective reality, and by means of these concepts we picture this reality to ourselves.
>
> In attempting to judge the success of a physical theory, we may ask ourselves two questions: (1) "Is the theory correct?" and (2) "Is the description given by the theory complete?" It is only in the case in which positive answers may be given to both of these questions, that the concepts of the theory may be said to be satisfactory. The correctness of the theory is judged by the degree of agreement between the conclusions of the theory and human experience. This experience, which alone enables us to make inferences about reality, in physics takes the form of experiment and measurement. (1935, 777)

Einstein and his collaborators define correct and complete from the point of view of physics: a baseball batted safely to left field is an instance of a force accelerating a mass, an event subject to measurement. That is its correct and complete description in physics. But not in baseball. In baseball, a correct and complete description focuses elsewhere, on the game itself: in baseball, we have a hit that puts the tying run on first. In other words, in the language of Ludwig Wittgenstein, correct and complete are relative to a particular language game, a particular form of life.

In this chapter, I want to illustrate Wittgenstein's distinction by offering two correct and complete descriptions of evolutionary taxonomy, the science of classifying animals and plants as species in accordance with evolutionary theory. One description will be from the point of view of science; another will be from the point of view of rhetoric. To evolutionary taxonomists, ours must be a world inhabited by evolutionary species. To each of these can be applied, from each of these can be inferred, at least in principle, a central process: the operation of natural selection on random genetic variation. Moreover, this reciprocal relationship between observation and theory presupposes objectivity: evolutionary taxonomy must describe a world whose existence is independent of its descriptions. In contrast, in a rhetorical description of evolutionary taxonomy, each aspect we have described as science must have its counterpart. *Sub specie rhetoricae*, we do not discover, we create: plants and animals are brought to life, raised to membership in a taxonomic group, and made to illustrate and generate evolutionary theory. To view evolutionary taxonomy in this way, it may seem, is to engage in the hermeneutics of suspicion, to place the objectivity at the center of the natural sciences in jeopardy, even to open the question whether these sciences have a right to exist. It might seem as though these two descriptions of the same phenomenon, the natural sciences, were rivals, competing for the same epistemic space. In this chapter, I hope to show the relationship between these two descriptions of science is not rivalry but complementarity.

In other words, the possibility of a complete rhetorical description has exactly the significance for rhetoric that a complete physical description has for physics or sociological description for sociology. It is no more and no less than the claim that rhetoric has disciplinary status, a claim as old as Aristotle in whose work the rhetorical five canons counted as a complete description of texts oriented toward persuasion. That no aspect of persuasion is omitted from Aristotle's rhetorical taxonomy does not exhaust the possibilities of persuasive speech as an object of analysis, the possibility of a grammatical description, example, or a description from the point of view of political science or psychology. These possibilities are complementary, not competitive.

The Scientific Description of Species

In contemporary biology, species cannot be defined classically by genus and differentia; the effort, in Wittgenstein's pungent phraseology, resembles trying to find the real artichoke by stripping it of its leaves. Instead, a species must be defined by depicting and describing "family resemblances" among organisms (1965, 125).

Earlier, in *Origin of Species*, Charles Darwin had placed this notion in a taxonomical context: "There are crustaceans at the opposite ends of the series, which have hardly a character in common; yet the species at both ends, from being plainly allied to others, and these to others, and so onwards, can be recognized as unequivocally belonging to this, and to no other class of the Articulata" (1964, 419).

When we define general terms like species according to family resemblances, we give up as spurious the precision of classical definition as it applies to the natural world. Nevertheless, definition by family resemblance is real in the sense that it marks off a nonarbitrary class. Because I lack rules that will demarcate absolutely what I name as a new species of hummingbird, I can be puzzled about whether a particular organism belongs to the species, but I can't just drop anything in there, eagles, for example. In addition, a good general term like hummingbird is an open class in two senses: there is always the possibility that I'll see a member I overlooked, and, anyway, at least for creatures not extinct, there is always the chance of genuinely new members. Finally, the application of a general term can be learned, that is, a learner can apply the term independent of the guidance of his teacher (Bambrough 1966). Let us see how this works by examining some typical papers in evolutionary taxonomy.

In "A New Species of Hummingbird from Peru" (Fitzpatrick, Willard, and Terborgh 1979), there is everywhere evident the extensive description and depiction that are the evidential grounding of family resemblance. These scientists, for example, record numerous qualitative visual impressions:

> A broad, pale buffy breast band separates the smaller throat spots from larger and more numerous discs on the breast and flanks. In a few specimens the posterior border of the breast band is entirely defined by a broad row of these discs. The belly is free of dark spots in all specimens. The downy crissum [anal region] is white as in males, and the undertail coverts [feathers at base] are dusky, edged Cinnamon. (1979, 178)

In addition to qualitative visual impressions, Fitzpatrick and his coworkers define the potential species by recording its behavior. For example, we are afforded a detailed description of "male-female display":

> Initially a pair [of birds] was foraging around the walls of [a vine-covered sinkhole] and in the surrounding shrubs along the rim. Both male and female were observed perching on a rootlet, making frequent sallies to capture tiny flying insects. (1979, 183)

To this abundance of qualitative information, quantitative data is added. We are given by degree, minute, and second the geographical coordinates of the area explored; to the nearest ten meters the height of the habitat above sea level; to the tenth of a millimeter the mean, range, and standard deviation of six specimen dimensions; to two decimal places the ratio of exposed culmen [upper ridge of a bird's bill] to wing chord length; to the second, the length of calls.

But the most impressive attempt to establish the species is in the frontispiece, a full-color painting of two of the birds in situ: a dimorphic pair, the female perched at an angle that most clearly displays her distinctive tail and underparts. The male displays not only his characteristic deeply cleft tail and dark coloring but his wings extended and in rapid movement as he hovers. Moreover, the bird's feeding behavior is shown: the "deep purple petals [of its favorite flower] form a tubular corolla that hangs vertically . . . forcing the foraging [bird] to hover directly below and point its bill straight upward to retrieve the nectar" (Fitzpatrick, Willard, and Terborgh 1979, 182).

The potential species is not only described and depicted in terms of family resemblances; by comparison and contrast, it is carefully differentiated from closely allied species. In this process of establishing species status, the character is the minimum unit of observation. Characters are parts of the organism or aspects of its behavior that most perspicuously establish it as a member of a genus or differentiate it as a separate species (Mayr 1982, 19–32). The harder a character works at either of these tasks, the more it is valued, the more heavily it is weighted. In establishing their potential species as a member of a particular genus, Fitzpatrick and his coworkers emphasize "the relative bill length, nostril feathering, and well delineated nasal operculum," a lid-like covering; in establishing the difference between the potential species and its fellows of the same genus, the authors give weight to the female's "elongated and deeply forked, entirely iridescent, metallic blue tail, equally bright on both surfaces; combined with buffy underparts interrupted by a pale pectoral [breast] band" (Fitzpatrick, Willard, and Terborgh 1979, 177, 181).

In the frontispiece and throughout their paper, pictures join words in the clear display of likenesses and differences, strongly reinforcing the presence of a potential new species. In figure 4.1, for example, the seven drawings considered in concert and without words make equally emphatic both the family resemblances crucial for genus identity and the contrastive features essential for within-genus differentiation.

It is by means of statistical inference that this extraordinary detail is given taxonomical significance: by this means, Fitzpatrick and his coworkers definitively assert that they have a new species, a new living entity related to all such entities. Sixteen specimens have been collected:

	Adult	**Subadult**
Male	5	5
Female	5	1

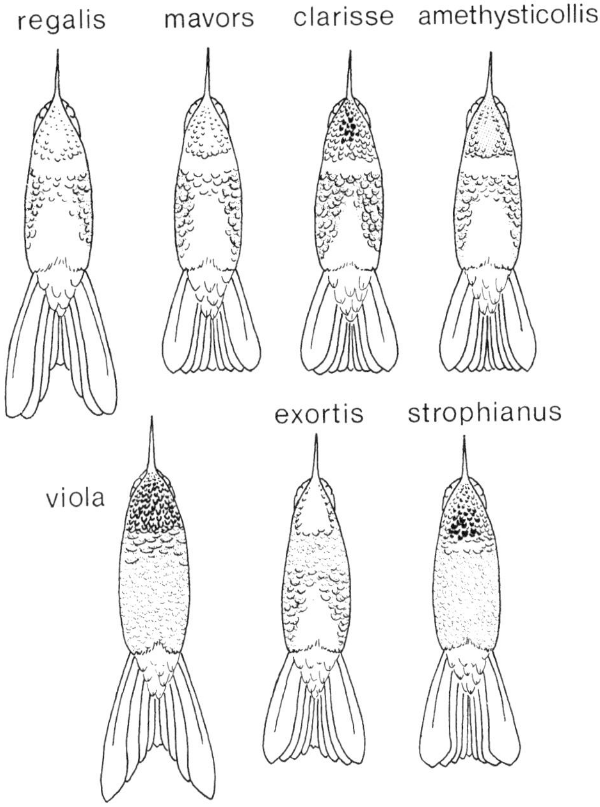

Fig. 4.1. Locating a New Species by Inspection. Ventral patterns and tail forms of *Heliangelus* females, including *H. regalis. H. micrastur* female resembles *exortis,* and *H. spencei* resembles *amethysticollis.* Note similarity between *regalis* and *mavors.* Reprinted figure from John W. Fitzpatrick, David E. Willard, and John W. Terborgh, "A New Species of Hummingbird from Peru," *Wilson Bulletin* 91 (1979): 180. With permission of Prof. John W. Fitzpatrick and the Wilson Ornithological Society.

Assuming a fair sample, the authors can describe the species by describing these specimens. Thus, when averages, ranges, and standard deviations are given, such as, those for wing-chord length, they are equally those of the specimens collected and of the species. Description, comparison, and statistical inference having done their work, the species can be officially certified as new by naming: it is a hummingbird, the genus *Heliangelus,* the species *regalis,* a Royal Sunangel.

But taxonomical identification is not the final stage: taxonomic species must be redefined in evolutionary terms. A species first characterized without Darwinian assumptions is now characterized in terms of a theory concerning the origin of species: that all stem from the operation of natural selection on random variations, some of which are favored because they increase the rate of survival of those who possess them.

In taxonomic work, evolutionary theory serves at least two purposes: it explains observations that arise in the process of identification, and it licenses prediction. In the process of identifying their new species of hummingbird, Fitzpatrick and his collaborators make the following problematic observation. Among the species of *Heliangelus* to which *regalis* seems most closely related, the monochromatic plumage of the male *regalis* is strikingly anomalous. To account for this irregularity, the investigators invoke the concept of convergence (similarity in species resulting not from descent but from environmental constraint):

> Thus, if *regalis* is indeed closest to this species as suggested by the appearance of the female, it has undergone a dramatic differentiation in which the male converged upon several more distant relatives. The elongated, narrow, metallic blue tail in both sexes of *regalis*, equally iridescent on both surfaces, is suggested only in *H. strophianus* (sexes similar) and *viola* (very large, dark green female lacks a breast band). (1979, 181)

But evolutionary theory does more than explain; it predicts. Prediction in evolutionary biology, of course, is not the same as in physics, where future events can often be forecast with startling quantitative accuracy. In evolutionary biology, although predictions need to be well-defined and consistent with the evidence on which they are based, they can legitimately vary in precision of measurement:

> For example, population ecologists are generally satisfied to explain to one order of magnitude the increases and decreases in population size of the organisms studied, and for this purpose net fecundity and mortality are usually sufficient. Game and fish management, however, may require prediction of population changes to an accuracy of 10 to 20 percent, and for this purpose complete age-specific mortality and fecundity schedules are required. Finally, the human demographer needs to project human population sizes to better than 1 percent accuracy, and to do so needs fecundity and mortality figures by age, sex, socioeconomic class, education, geographical location, and so on. (Lewontin 1984, 7–8)

Lewontin's principle is implicit in a report by A. de Ricqlès and J. R. Bolt. These workers inquire into the origins and nature of "perhaps the most important captorhinid [prehistoric reptile] character," a feature whose evolutionary history they regard as a prerequisite to proper taxonomic use: jaw growth and tooth replacement (1983, 7). Their predictions, though far from precise, are nevertheless legitimate because their criteria are well-defined and consistently applied. On the basis of large, clearly visible anatomical features, these scientists create a preliminary model of jaw growth and tooth replacement. For verification, they check their predictions against effects of jaw growth and tooth replacement not visible to the naked eye. It makes sense that the dynamics of these processes will leave traces in surviving fossils, and that, if the model is correct, effects visible to the naked eye will be consistent with effects visible at between two and six magnifications.

Among the papers in evolutionary biology I have examined, falsifiability is mentioned almost as much as prediction. Falsifiability is prediction turned on its head; it asserts that a theory can be undermined by its repeated failure to predict. This notion has Sir Karl R. Popper as its best-known exponent (1965; 1968; see also Medawar 1984). For Popper, the possibility of falsifiability is at the root of all good science, all of which is open to being "*refuted by experience*" (emphasis in original). In other words, it is the essence of science that we can refute theory with contravening facts:

> It is possible by means of purely deductive inferences (with the help of the *modus tollens* of classical logic) to argue from the truth of singular statements to the falsity of universal statements [in the texts examined, from observations to hypotheses or models]. Such an argument to the falsity of universal statements is the only strictly deductive kind of inference that proceeds, as it were, in the "inductive direction"; that is, from singular to universal statements. (1968, 41)

Popper uses the term *modus tollens* to refer to the rule of logic that, although no number of facts, no matter how great, can confirm a theory, a single fact if true can disconfirm it. As an example, let us take the theory of evolution and the alleged fact that the world is 6,006 years old. Both the theory and the fact cannot be true. So if the fact is true, the theory must be false. Popper recognizes that theories under attack from uncomfortable facts can always be saved by introducing an auxiliary hypothesis or by changing a definition merely because it is convenient to do so. His solution is "the empirical method" with "its manner of exposing to falsification, in every conceivable way, the system to be tested" (1968, 41–42).

Two papers in my sample claim an openness to falsifiability. Writing about the spread of chipmunk populations, B. D. Patterson says of an hypothesis integral to his model that it is "apparently refuted by the presence of *E. canipes*, rather than *E. quadrivittatus* [species of chipmunk], in the Gallinas mountains"; according to this model, "small mountain ranges isolated from centers of endemism [areas where species originate] should have been colonized by southwardly advancing taxa," of which *E. quadrivittatus* is one (1982, 393–94). In their paper, Ricqlès and Bolt assert that "evidence for successive occupation by several generations of teeth at the same site would falsify [their] model" of reptilian jaw growth and tooth replacement, a model that implies "that in the MR [multiple-rowed] area a given attachment site was occupied only once" (1983, 22). Because the geography of Arizona and New Mexico allows for the apparently anomalous presence of *E. canipes* in the Gallinas mountains, Patterson's hypothesis survives; because successive same-site tooth occupation is not in evidence, the conclusion of Ricqlès and Bolt still rests on firm ground.

The analysis of the concept of species by means of an examination of typical papers in evolutionary taxonomy supports the view that science is an inductive progress to reliable knowledge, knowledge always open to, but nevertheless so far resistant to, falsification. The induction from which this knowledge proceeds has observational data as its raw material, data that do not depend at all on the evolutionary theory that explains them. For example, the collection of sixteen similar birds by Fitzpatrick and his coworkers is a task possible to earlier, evolution-free centuries. It need have been motivated by nothing more than the hunch that the birds might form a new group. Of course, once the group has actually been segregated, it can be defined taxonomically; and, once it has been so defined, it can be redefined according to evolutionary theory. Evolutionary theory can then be used to explain and predict events and entities in the natural world and to retrodict these in the fossil record.

The Rhetorical Description of Species

According to Chaim Perelman and Lucie Olbrechts-Tyteca in their *New Rhetoric*, scientists form a "universal audience." Despite their small numbers, such audiences are universal in the sense "that everyone with the same training, qualifications, and information would reach the same conclusions" (1971, 34). This does not mean that any actual audience for a particular science possesses these characteristics. The universal audience is simply one that must be presupposed; it is the audience that scientists must see themselves as addressing when they write or speak. It is by means of this universal audience that the natural sciences come within the sphere of rhetoric; their rhetorical description would characterize "the image [the evolutionary taxonomist] holds of the universal audience he is trying to win over to his view" (1971, 33). To accomplish this task, rhetorical equivalents for the stages of the scientific description we have just presented must be found: we must show how scientists create a world that persuades their fellows, a world in which plants and animals have been brought to life, raised to membership in a taxonomical group, and made to illustrate and generate evolutionary theory.

Potential species are brought to life by giving them presence, by placing these purposeful collections of living things at stage center in audience consciousness (Perelman and Olbrechts-Tyteca 1971, 142, 116–17). In evolutionary taxonomy, this presence is created chiefly by the use of multiple sensory perspectives combined with overdescription, the characterization of sense objects in detail far beyond an ordinary reader's expectations. In bringing to life a new avian species, for example, Fitzpatrick and his coworkers are no purists looking for essential differences: all descriptive features are mustered. This multiplicity of perspectives creates presence so effectively because it mimics the methods by which we confirm our everyday impressions of the reality of ordinary objects. We supplement visual data with more visual data from different angles and distances; the data of one sense with the data of another; qualitative sense data with mensuration. Thus, the presence of a species increases as their paper progresses; until the "whole field of consciousness" becomes filled with this creature "so as to isolate it, as it were, from the [audience's] overall mentality" (Perelman and Olbrechts-Tyteca 1971, 118).

In scientific, as distinct from literary prose, the resources of language from which presence is created are decidedly limited. Because scientific prose is designed to create the impression that its language refers unproblematically to a real world existing independently of any perceiving subject, it generally excludes the subjective dimension of description, the use of emotion-charged words or irony. For the same reason, scientific prose generally excepts any device that shifts the reader's attention from the world that language creates to language itself as a resource for creating worlds. These restrictions account for the unavailability of many of the techniques of presence discussed in *The New Rhetoric*: punning would be just one example of such an exclusion.

But language so constrained is hardly free from rhetorical effect; as Roland Barthes cogently observes: "the zero degree . . . *is a significant absence*. . . . the absence of rhetorical signifiers constitutes in its turn a stylistic signifier"; "denotation is . . . the *last* of the connotations . . . the superior myth by which the text pretends to return . . . to the nature of language, to language as nature" (1968, 77–78; 1974, 9; emphasis in original). Because the creation of presence in science is limited to those devices by which language may be said to refer unproblematically to the world, it seems fair to give that sort of presence a name: to call it *referential presence*.

We can account for referential presence *sub specie rhetoricae*; but can we account for its taxonomic transformation? We can if we can show how statistical inference, naming, and artistic rendering create the persuasive structures that transform referential into taxonomical presence.

In evolutionary taxonomy, statistical inference is designed to convince readers that speciation partakes of the certainty generally attributed to a branch of mathematics. There is no question of the importance to evolutionary taxonomy of this routinely used set of techniques, but it is far from clear that statistical inference is the actual basis for the establishment of new taxa. The validity of such inferences depends on the efficacy of sampling procedures: generalizations from a sample can be made only on the basis of random selection from a defined population. In practice, limitations of time and funding force reasonable compromises, but Fitzpatrick and his coworkers define *Heliangelus regalis* by means of sixteen birds. How can they assert that these are representative? This use of a sample size that cannot be statistically significant, which seems the taxonomic norm (Sokal and Crovello 1984, 544, 558), provides a clue to the rhetorical nature of the taxonomic transformation.

In truth, taxonomic speciation depends less on statistical inference than on a fundamental presupposition about the existence of species in the order of nature. These species must necessarily be imagined as "discrete sets with appreciable distances between them." Taxonomic speciation absolutely "depends upon the possibility of grouping data in clusters with empty space between them" (Kuhn 1977, 312; 312n.). This continuity within and discontinuity between species are perfectly in accord with gestalt psychology: between gestalten there is "a 'dead' interval which corresponds . . . to the mere extension or ground outside a visual shape" (Köhler 1947, 111). However, this presupposition is not in accord with the conviction that species are established solely on the basis of statistically inferred regularities.

A striking instance of the operation of the presupposition concerning the essential nature of species, culled from the papers I examined, concerns *Trachelyichthys exilis*—a new species of a new genus of catfish identified and described by D. W. Greenfield and G. S. Glodek on the basis of a single specimen! While statistical inference is invoked, a different and antithetical procedure is actually used. Greenfield and Glodek simply assume that a single creature can evoke the order of nature as a taxonomic network. Forewarned by such imaginative leaps, one may question the legitimacy with which the precision demonstrated for specimens is routinely transferred to the species.

Artistic rendering also confers taxonomic identity. In Fitzpatrick, Willard, and Terborgh on hummingbirds, the full-color portrait of *regalis* in situ may strike the naive viewer as a candid glimpse. But this picture is clearly posed: it presents these creatures in a manner designed to display not just any characters but only those that best distinguish them from their fellows. In the case of the line drawing reproduced in figure 4.1, the purposes of taxonomic identification are served by another form of visual rhetoric: simplification of line and shade in the interest of taxonomic identification.

Finally, naming confers taxonomic identity:

> Wherever language and men exist, there is not only a freedom from the pressure of the world, but this freedom from the habitat is also freedom in relation to the names that we give things, as stated in the profound account in Genesis, according to which Adam received from God the authority to name creatures. (Gadamer 1975, 402)

The verbal portrait of the hummingbird consists of scattered islands of detail. Naming alone resolves the paradox by means of which this scattered detail becomes a new creature, one for whom a taxonomic space already exists (Barthes 1974, 60–63, 94–95, 209–10).

The final stage is the transformation of the taxonomic species to the evolutionary species. The evolutionary species is part of at least two networks of meaning, two versions of evolutionary theory. In the first, or strong, version, the species is an integral part of a formulation that aspires to make quantitatively precise forecasts of events in space-time, events directly inferred from mathematically expressed physical laws. In the second, or weak, version, this same concept is an integral part of a formulation that brings together a large and otherwise disparate number of phenomena in the natural world under a single conceptual umbrella.

Darwin himself is the source of both versions of evolutionary theory. In *Origin*, taking his cue from the explanatory model of classical physics, he says in support of the strong version: "Throw up a handful of feathers, and all must fall to the ground according to definite laws; but how simple is the problem where each shall fall compared to that of the action and reaction of the innumerable plants and animals which have determined, in the course of centuries, the proportional numbers and kinds of trees now growing on the old Indian ruins!" (1872, 86). At other times, however, in his letters, Darwin seems to have for evolutionary theory explanatory goals far less rigorous: "An hypothesis is *developed* into a theory solely by explaining an ample lot of facts"; it connects these facts "under an intelligible point of view" (1959, 80; 210; emphasis in original). In a letter to his friend Charles Lyell, Darwin specifically abjures strong prediction, which he calls "prophecy" (1959, 9–10). Indeed, as a whole, *Origin* exemplifies the weak version of evolutionary theory, but to adopt this version is also to abjure falsifiability as a criterion for the adequacy of theoretical formulations.

Are evolutionary taxonomists strong or weak theorists? Weak, for the most part. But in the papers examined, Ricqlès and Bolt and B. D. Patterson clearly underwrite falsifiability as the chief criterion for the theoretical adequacy of their formulations. In so doing, they seem to accept the evolutionary species as part of a strong theory. However, appearances are deceptive. In his analysis of falsifiability, Popper insists that the element of risk be maximized with "an attempt to solve an interesting problem by a bold conjecture, *even (and especially) if it soon turns out to be false*" (1965, 231; emphasis in original). Although the statements of Ricqlès and Bolt look Popperian at first glance, careful scrutiny reveals the hyperbolic nature of their claims. In these papers, the theories of their authors are not seriously at risk; consequently, evolutionary theory is never seriously at risk.

Ricqlès, Bolt, and Patterson take advantage of the impressiveness that an openness to falsification confers without taking the risks that such openness should entail. What Ricqlès and Bolt identify as falsifiable is their preliminary model, a model that is no more than an artifact of their method. They elected to build this model from large, clearly visible anatomical features and to correct it by means of anatomical features not visible to the naked eye. As they readily admit, they could have built their final model directly and entirely from submacroscopic features. If they had, potential falsifiability would have disappeared, because submacroscopic evidence "for successive occupation by several generations of teeth at the same site" would have been clearly absent from the beginning (1983, 22).

Patterson is no truer to Popper than are Ricqlès and Bolt. Because Patterson's model really is endangered by some awkward evidence, it is technically falsifiable. But, to save it, he resorts to a hypothesis of convenience, expressly forbidden by Popper: "An elevated corridor . . . extends from the White Mountains north to the Gallinas Range. Chipmunks *evidently* were able to move through this corridor to colonize the Gallinas Range before it could be reached by *E. quadrivittatus*" (1982, 395; my emphasis). Falsifiability is invoked in both these instances to give an impression of the strength of particular taxonomical claims.

Despite differences, the strong and weak versions of evolutionary theory are importantly similar. In both, natural selection creates the phylogenies, the lines of descent we observe in nature and in the fossil record; in both, evolution is a theory of process and a theory of descent. It is these shared theoretical resources that allow taxonomists to use either version of evolutionary theory as a completely unquestioned, utterly reliable source of intellectual underpinning, an ideational account on which they can reliably draw to explain and extend their observations. Because evolutionary theory founds evolutionary biology, by definition, evolutionary biology has axiomatic status: in every paper, its appropriateness in species identification can be taken as settled.

It is far from settled. When Sokal and Crovello assert that taxonomy can proceed very well without the evolutionary species, "a theoretical ideal to which existing situations are forced to fit as closely as possible" (1984, 560), they are not alone, nor is their alternate approach—phenetics—without adherents. Nor is phenetics the only alternative. An advocate of "transformed cladistics" believes "that much of today's explanation of nature, in terms of neo-Darwinism or the synthetic theory, may be empty rhetoric" (C. Patterson 1982, 119).

Although this last accusation may itself be dismissed as empty rhetoric, it ought not to be. It is symptomatic of a general intellectual problem: the paradox inherent in any affirmation of evolutionary theory. If the theory is right, species cannot be natural kinds, entities with atemporal identities, like quarks or electrons: full knowledge of intermediary varieties would demonstrate their wholly historical nature. Concerning Ernst Mayr's classic *Systematics and the Origin of Species*, Niles Eldredge says: "His treatment of the entire problem implies a recognition that if one goes too far in embracing the principles of natural selection and adaptation as the be-all of evolution, the systematist [taxonomist] is left with nothing to explain" (1982, xix). In other words, any version of evolutionary theory eventually leads to the disappearance of the species as a legitimate natural kind. The evolutionary species is an oxymoron created only by avoiding the full implications of the theory on which its existence apparently depends.

Complementary Interests, Complementary Descriptions

These two descriptions of evolutionary taxonomy—the scientific and the rhetorical—correspond to the first two categories in Jürgen Habermas's taxonomy of human interests. The first interest is characteristic of the natural sciences; it is the "cognitive interest in technical control over objectified processes" (1971, 309). While an analysis of this interest reveals the conditions for the rise of natural sciences in the modern era, it tells us nothing about the place of these sciences in the conduct of life. Questions concerning this place are the province of Habermas's second interest, the practical cognitive, whose task is the creation and comprehension of meaning:

> Hermeneutic inquiry discloses reality subject to a constitutive interest in the preservation and expansion of the intersubjectivity of possible action-orienting mutual understanding. The understanding of meaning is directed in its very structure toward the attainment of possible consensus among actors in the framework of a self-understanding derived from tradition. (1971, 310)

In other words, the practical cognitive interest creates an interdependent social and political world founded on mutual understanding. This is a world in which the rhetorical perspective I have been exploring has a central role.

The relationships between technical and practical interests and between science and rhetoric are complementary: without the technical interest, science is impossible; without the practical, science would be devoid of meaning for the conduct of life (Habermas 1971, 309–11; see also 195, 212). In the case of the practical interest, what might seem at first derogatory, cutting science down to size is, on reflection, merely deflationary, bringing science down to earth, acknowledging that scientists are also social and political actors.

While the practical interest makes the social and political world possible, it does not free this world from the influence of the "illusion of objectivism" cultivated by the sciences and such fellow travelers as exist in their long shadow—disciplinary formations such as clinical medicine and psychiatry, on the one hand, and food and consumer science on the other. This is an illusion detrimental to human flourishing to the extent that it is designed to exclude other legitimate perspectives. To prevent this unhealthy colonization, we have Habermas's third interest, the emancipatory. Its purpose is to destroy the illusion of objectivity "through demonstrating what it conceals: the connection of knowledge and interest" (1971, 316–17). Chapter 10 discusses how the "technical control of objectified processes" works against human flourishing by concealing the clash between the public good and private interests.

5 The Possibility of a Complete Rhetorical Description of Scientific Change

Newton's *Opticks* as Example

In science, there are two sorts of revolutionary masterpieces: those rhetorically powerful enough to provoke open revolt and those rhetorically ingenious enough to avoid it. The first employs a line of argument, or *topos*, of complete rupture with the past, of radical dismissal of its mistaken views; the second employs the *topos* of essential continuity, the perfection of past views that were not so much incorrect as incomplete. *Dialogue Concerning the Two Chief World Systems* and *On the Origin of Species* are examples of masterpieces of the first sort. Initially, Galileo and Charles Darwin caused more debate than assent, more turmoil than change. René Descartes's *Dioptrics* and Isaac Newton's *Opticks*, on the other hand, are masterpieces of the second sort: each in its own way successfully persuaded; each dominated optical research for nearly a century. In these works, each man argued for change less as revolution than as continuity: the extension of the best of the past.

Published in 1637 as an integral, though now nearly forgotten, appendix to his *Discourse on Method*, Descartes's *Dioptrics* has been quite properly called a first step in a new direction. Nevertheless, Descartes did not violate traditional presuppositions. Despite the newness of his program, he shared with traditional science the conviction that white light was basic and that color was derivative, an alteration of white light. He also believed with tradition that rational intuition, not experiment, was prior in the order of knowledge: reason, not experience, was the bedrock, the touchstone, of knowledge. Finally, he was as convinced as Aristotle that a complete scientific explanation must include at least the traditional three causes: the formal, the efficient, and the material. These shared convictions allowed Descartes to be rhetorically transparent and convincing at the same time: he could use his prose to clarify, indeed to highlight, his views.

In 1672, a time when Descartes's optics was already firmly entrenched, Newton published his first paper on that science. Its views were new in a new way. Unlike Descartes, Newton challenged a traditional tenet concerning the nature of light. For the first time in the history of optics, white light was revealed as derivative, as a compound of all of the lights of the visible spectrum. Moreover, the ground Newton offered for certainty in optics differed radically from that of his predecessor and of his predecessor's predecessors: by giving priority in the order of knowledge to experiment over rational intuition, Newton overturned a central presupposition of traditional science. In addition, Newton's explanation of light did not give a full account of its origin, an account that included the operation of its material cause. A startling claim, a new method, a different, more restrictive style of explanation—seemingly, Newton needed to discharge a strong burden of proof. But in this early paper on light and color, in a rhetoric as transparent as that of Descartes, Newton did not discharge this burden; instead, he emphasized his conflict with traditional views and methods.

After a flurry of inconclusive debate and a second early paper, Newton withheld his optical research from the public for nearly three decades. In 1704, he published his *Opticks*, his second attempt at persuasion. In this work, Newton discarded the transparent rhetoric that made the epistemological and explanatory novelty of his early papers clear; in its stead, he substituted a rhetoric that invented an essential continuity between his work and the optical and scientific past. The rhetoric of *Opticks* concealed his radical intent; it was designed to convince at the expense of frankness. In his final masterpiece, Newton transformed optics, and experimental science, by allowing his readers to believe that an adherence to the new did not entail a fundamental rejection of the old. This strategy was successful: throughout the eighteenth century in England and on the Continent, the physics of light was Newton's physics.

For any advocate of the complete rhetorical description of science, the story of optics from Descartes to Newton constitutes an apparent stumbling block. While a full rhetorical description may be available for the *Dioptrics* and for Newton's first paper on optics, the transition from the first to the second turns on an empirical result seemingly outside the realm of rhetoric. In addition, although Newton's early paper and his later *Opticks* may each be susceptible to a full rhetorical description, the transition from one to the other seems to depend on the survival of a common core of meaning, a core also seemingly outside the reach of rhetoric. In this chapter, after an analysis of Descartes's and Newton's work that is, I believe, fully rhetorical, I address the two transitions—that between Descartes and Newton and that between early and late Newton—and argue that the case for full rhetoricity holds not only for the works in question but also for the transitions between them. For those who are convinced by these arguments, there will be no province of science immune from rhetorical description; for those who remain unconvinced, a border will have been drawn beyond which rhetoric must not trespass.

Cartesian Optics

Fully in tune with tradition, Descartes agreed that white light was basic, and color derivative, that is, an alteration in white light. He also concurred with tradition in representing light as straight lines that can cross without interfering. This characteristic permitted him to use geometry to solve problems in the physics of light: the axioms and theorems of Euclidian geometry are central to the methodology of both traditional and Cartesian optics.

To illustrate the two central geometrical properties of sunlight—its rectilinear propagation and the noninterference of its rays—Descartes compares light's action to the pressure on a vat of half-pressed grapes. This pressure transmits itself throughout the vat, squeezing the juice equally through two widely separated holes in the bottom:

> And in the same way considering that it is not so much the movement as the action of luminous bodies that must be taken for their light, you must judge that the rays of this light are nothing else but the lines along which this action tends. So that there is an infinity of such rays which come from all points of luminous bodies, toward all points of those that they illuminate, in such a manner that you can imagine an infinity of straight lines, along which the actions coming from all points of the surface of the wine . . . tend toward [one hole], and another infinity, along which the actions coming from these same points tend also toward [the other hole], without either impeding the other. (Descartes 1965, 70)

When geometric methods are applied, light is found to exhibit two regularities: reflection and refraction. When a ray of light is reflected, its angle of incidence, the angle at which it strikes the reflecting surface, is always equal to the angle of reflection, the angle at which the ray leaves the reflecting surface (fig. 5.1).

When light penetrates a transparent body at an angle other than right, it refracts; it bends toward or away from an imaginary line called the normal, a line perpendicular to the surface. When the light crosses into a medium of higher optical density, for example from air into water, it bends toward the normal; when it crosses into a medium of lower optical density, it bends away (fig. 5.2).

The angles of incidence and reflection are equal, a quantitative relationship known since antiquity. An analogous quantitative relationship for refraction eluded all investigators until Willebrord Snel and Descartes. In refraction, as both

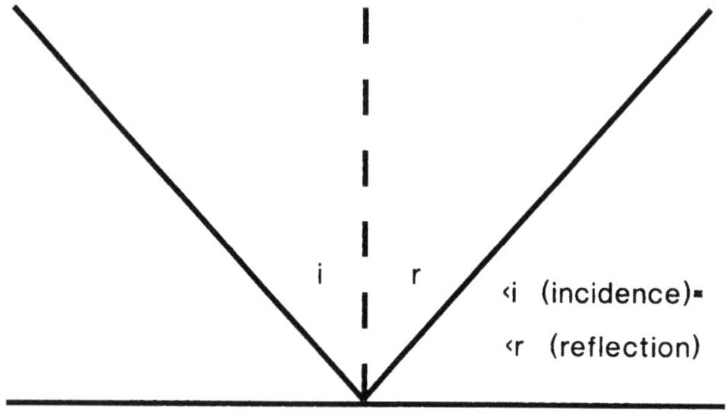

Fig. 5.1. Reflection

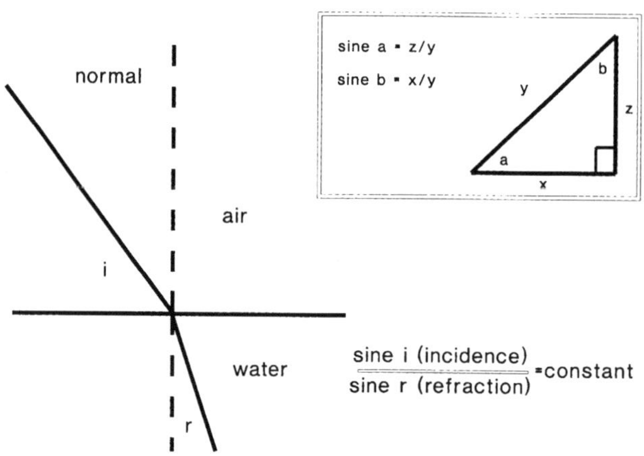

Fig. 5.2. Refraction

discovered, rays of light conform to a definite geometrical relationship, that of sines. In a triangle with a right angle, the sine is the ratio of the side opposite either acute angle and the side opposite the right angle. A ray refracting in most transparent substances bends so that the ratio of the sine of its angle of incidence to the sine of its angle of refraction equals a specific number. For each regularity-abiding refracting substance, this number never changes:

$$\frac{\text{sine i}}{\text{sine r}} = \text{constant}$$

Although the sine relationship was a new discovery, it was an empirical regularity compatible with traditional assumptions and methods: there was no question of its immediate absorption into the body of traditional optics.

While Descartes practiced traditional geometric optics, he did so within the framework of the new science he created. Scientists working within traditional optics might easily be persuaded that the refractive regularity of Snel and Descartes was not merely a useful empirical regularity but a law of science; wrongly persuaded, Descartes's view. To Descartes, an empirical regularity becomes a scientific law only when it is understood as an integral part of a true and coherent physics, his physics. The refractive regularity can be promoted to a law only when it is understood as correct according to the central and absolutely certain principles of Cartesian physics.

Descartes shares with tradition his conviction that the goal of science is absolute certainty rather than moral certainty (what we would call very high probability). However, the source of his certainty represents a radical interpretation of the primacy of reason in the order of knowledge, one that leads to a new, counterintuitive physics. Provided they are clear and distinct, the intuitions of Descartes's unaided reason form the incorrigible foundation of his science. According to these, the material world consists only of extension and its laws of motion. The universe is a plenum, completely full: space itself is material, because extended. Local motion, the relative repositioning of extended substances, can occur only through contact. Everything we experience, even light, is the result of local motion originating in contact.

In Cartesian physics, the sun consists of small, rapidly moving particles of the first element that exert a pressure on the second, the element that fills interplanetary spaces. This pressure is transmitted instantaneously in straight lines in all directions. Experienced on the retina, this pressure creates the sensation called light. Light *is* this pressure; it is not, as it was for Aristotle, a qualitative alteration in a medium, the air; it is not, as it will be for Newton, the effect on the retina of a stream of tiny particles traveling from a luminous source.

At the end of his *Principles of Philosophy*, Descartes sums up the full implications of his position:

> These reasonings of ours will perhaps be included among the number of these absolutely certain things by those who consider how they have been deduced in a continuous series from the first and simplest principles of human knowledge. Especially if they sufficiently understand that we can feel no external objects unless some local movement is excited by them in our nerves; and that such movement cannot be excited by the fixed stars, very far distant from here, unless some

> movement also occurs in these and the whole intermediate heaven; for once these things have been accepted, it will scarcely seem possible for all the rest, at least the more general things which I have written about the World and the Earth, to be understood otherwise than as I have explained them. (1983–84, 287–88)

In Descartes's physics, then, reason is sole guarantor of the certainty of the general principles from which the specific truths of this science are derived. In a revealing aside on the relationship between reason and experiment, Descartes asserts: "And the demonstrations of this [law of impact] are so certain that, even if experience were to appear to show us the opposite, we would nevertheless be obliged to place more trust in our reason than in our senses" (1983–84, 69n.). It is our reason alone that assures us: no explanation of light, no explanation in physics can be persuasive if it is not also mechanical, the result direct or indirect of matter in motion: "Whatever I concluded to be possible from the principles of my [natural] philosophy actually happens whenever the appropriate agents are applied to the appropriate matter" (1981, 38).

Like light, color must be explained in terms of matter in motion. Recall that in Descartes's view, the particles of the second element, those that fill all the space between the sun and the earth, transmit the sun's first-element pressure. In refraction, these second-element particles take on differing rotational speeds, speeds whose pressures are communicated directly to the retina. It is from these differing pressures that the sensations of color arise: "The nature of the colors appearing at [the lower end of the spectrum] consists only in the fact that the particles of the fine substance that transmits the action of the light have a stronger tendency to rotate than to move in a straight line; so that those which have a much stronger tendency to rotate cause the color red" (1965, 337). The sensation of white light is the result of the simple pressure of particles of the second element on the retina; color is the result of the action of these same particles to which a rotating motion has been imparted.

Although reason is central to Cartesian physics, experience cannot be ignored; indeed, observation and experimentation have an important role. Though everything flows from general principles, everything cannot be deduced from them: the constant of optical density that accounts for the degree of refraction differs for differing substances and cannot be deduced from the law of refraction. To obtain this constant in each particular case, "we must appeal to experience" (1965, 81). Even the law of refraction, though it conforms to the laws of motion, cannot be deduced from them. Light always acts in accordance with the laws of motion, which are general principles, but it does not always act in accordance with the law of refraction. Experience is the sole desideratum for empirical regularities:

"For, seeing that these parts could have been regulated by God in an infinity of diverse ways, experience alone should teach us which of all these ways He chose" (1983–84, 106).

An examination of Descartes's philosophy, it is true, reveals a role for observation and experiment that is more nearly central. In that philosophy, reason tells us that the real qualities of material objects with which science is concerned consist "clearly and distinctly" of "magnitude or extension in length, breadth, or depth" (1983, 164; see also 56–57; 1983–84, 76–77). Such a view seems to entail a quantitative physics, but neither mathematics nor measurement appears regularly in Descartes's science, pages filled with accounts not susceptible to measurement, accounts no experiment could ever confirm. Despite his statements to the contrary, Descartes's physics seems essentialist and qualitative (cf. Gaukroger 1980, 134).

In all of Descartes, there is one exception to this subordinate role for observation in the order of knowledge, an exception that proves this rule. In his discussion of the rainbow, Descartes signally advances the cause not only of his but of our physics; by means of observation and measurement, he deduces the radii of both rainbow arcs: "The radius of the interior arc must not be greater than 42°, nor that of the exterior one smaller than 51°" (1965, 339). *Must be*! From Descartes's discourse on the rainbow, which Newton knew well (1969, 543–49), one could learn a whole new way of doing science, a way in which observation and measurement combined to form a vanguard in search of new, durable empirical regularities and new laws of science.

However, from this analysis, which he regarded as an authentic example of his method (1981, 46), Descartes learned nothing new about the fundamental role of observation and measurement: he gives equal credence to his theory of the arc of the bow, amply confirmed by observation and measurement, and to his theory of color, for which he offers, in addition to a general conformity with experience, only the certainty of his unique, reason-derived mechanics (1965, 338). Concerning this latter certainty, Descartes sounds a characteristic note when he asserts "how little faith we must have in observations which are not accompanied by true reason" (1965, 342).

Newton's Optical Work: His First Paper

In his first paper on optics, published in the *Philosophical Transactions* of 1672, Newton tells the story of his discovery:

> In the beginning of the Year 1666 . . . I procured me a Triangular glass-Prisme, to try therewith the celebrated *Phaenomena* of *Colours*. And in order thereto having darkened my chamber, and made a small hole in my window-shuts, to let in a convenient quantity of the Suns

> light, I placed my Prisme at his entrance, that it might be thereby refracted to the opposite wall. It was at first a very pleasing divertisement, to view the vivid and intense colours produced thereby; but after a while applying my self to consider them more circumspectly, I became surprised to see them in an *oblong* form; which, according to the received laws of Refraction, I expected should have been *circular*. (1978, 47–48)

As Newton tells his story, the anomaly of the oblong spectrum demands an explanation: if light behaved as Descartes said, the spectrum would disperse equally in all directions, like spray from a hose. In his search for a cause of this anomaly, Newton discards explanation after explanation: perhaps it was the thickness of the prism glass, perhaps its unevenness; perhaps the rays' curve. All of these "suspicions" being removed, Newton performs the "*Experimentum Crucis,*" the decisive experiment; the final searching question is put to nature (1978, 50).

To this question, the optical tradition has no answer: it exhibits a fatal flaw, a deep-seated but hitherto unsuspected incoherence. Unless we abandon the law of sines—an unthinkable prospect—we must abandon the belief that white light is basic, color derivative; we must believe instead that color is basic, white light derivative. The crucial experiment clearly and unequivocally shows that white "*Light* consists of *Rays differently refrangible,*" that is, bendable (1978, 51). When these rays are separated, they form the radiant colors of the spectrum; combined again, in the proper proportions, they recreate white light.

In thirteen propositions that follow the *Experimentum Crucis*, Newton shows how the differing refrangibilities of light explain "the celebrated *Phaenomena* of *Colours*" (1978, 47). Descartes's views notwithstanding, sunlight is "ever compounded, and to its composition are requisite all the aforesaid primary Colours, mixed in a due proportion" (1978, 55). It follows from the combinatory nature of white light that "Colours are not *Qualifications of Light*, derived from Refractions, or Reflections of natural Bodies [as in Descartes] . . . but *Original* and *connate properties*, which in diverse Rays are divers" (1978, 53). This characteristic of light explains not only radiant colors (those produced by a prism) but "the Colours of all natural Bodies [which] have no other origin than this, that they are variously qualified to reflect one sort of light in greater plenty then another" (1978, 56).

It is a deliberate irony that the spearhead of Newton's attack on Descartes is the certainty of an anomaly founded on experience: it will not go away; it cannot be explained away. It is equally ironic that only a wholly quantitative point of view can reveal the anomaly: unlike Descartes, Newton projects his spectrum not a few inches, where the anomaly is not apparent, but twenty-two feet, where it is unmistakable (Westfall 1984, 164). Like Johannes Kepler's famous eight minutes of arc—a discrepancy that led to his famous laws of planetary motion—these twenty-two feet have revolutionary implications in Newton's hands and, in his early paper, he makes the most of them.

In this paper, following in the footsteps of Robert Boyle and Robert Hooke, Newton uses experiment systematically as the primary instrument in the search for knowledge. However, Newton's relentless use represents a new departure, as a comparison with Boyle's *Experiments and Considerations Touching Colours* and with Hooke's *Micrographia* will confirm. In theory and in practice, Newton clearly and uncompromisingly reverses the traditional and Cartesian roles of reason and experiment: "If the Experiments, which I urge, be defective, it cannot be difficult to show the defects; but if valid, then by proving the Theory they must render all Objections invalid" (1978, 94). In Newton's science, experiment bears the full weight both of discovery and of theory.

In a justly famous Baconian pronouncement on method, Newton affirms experimentation as the driving force behind theory and as the prime source of certainty. In the same passage, he discloses his perception of the central defect of Cartesian physics, indeed of any physics that overemphasizes the hypothetical—the fatal ease with which explanations may be invented:

> The best and safest way of philosophizing [doing science] seems to be this: first to search carefully for the properties of things, establishing them by experiments, and then more warily to assert any explanatory hypotheses. For hypotheses should be fitted to the properties which call for explanation, and not be made use of for determining them, except in so far as they can furnish experiments. And if anyone makes a guess at the truth of things by starting from the mere possibility of hypotheses, [I] do not see how to determine any certainty in any science; if indeed it be permissible to think up more and more hypotheses, which will be seen to raise new difficulties. (1959, 1:169)

Newton's first paper failed to convince. His notion of a crucial experiment, one whose result is proof against argument, is both central to his case and seriously flawed as a rhetorical device. The persuasiveness of the crucial experiment depends on its replicability; but the crucial experiment in this first paper is accompanied by neither diagram nor clear directions. More significant because far less remediable, the persuasive effect even of replicable results depends absolutely on their univocal interpretation. At one point, Newton becomes exasperated with some rival experimental results: "It is not the number of Experiments," Newton cries, "but weight to be regarded; and where one will do, what need many?" (1978,

174). But it was folly to expect such unanimity of interpretation in so hotly disputed a seventeenth-century subject as the nature and action of light. In his first paper, Newton draws what for him is the obvious inference that light consists of rays composed of particulate matter: "Since Colours are the *qualities* of Light, having its Rays for their intire and immediate subject, how can we think those Rays *qualities* also, unless one quality may be the subject of and sustain another; which in effect is to call it *Substance*" (1978, 57). But Hooke and the French scientist Ignace Pardies had different though plausible explanations of Newton's experimental result, explanations founded on the premise that light consists not of rays but of waves.

Critics also objected to the incompleteness of Newton's explanations. The Dutch scientist Christiaan Huygens animadverted that "he hath not taught us what it is wherein consists the nature and difference of Colours, but only this accident . . . of their *different Refrangibility*" (Newton 1978, 136). Indeed, Newton had avoided any mechanical explanation of refraction or color formation: "I shall not mingle conjectures with certainties," he grandly asserted (1978, 57). But to Huygens, a scientific explanation that insisted on the corpuscularity of light must include a specific mechanism for the production of the sensations of light and color by means of matter in motion.

Newton's 1672 paper succeeded in establishing neither his method of doing science nor his beliefs about the nature and action of light. In his *Opticks* of 1704, Newton created a second opportunity to persuade the scientific community of the efficacy of his method and the truth of his optical beliefs, the same method and, essentially, the same beliefs contained in his optical papers of thirty years before. It is to the *Opticks*, Newton's final masterpiece, that I now turn.

Newton's *Opticks:* A Rhetorical Masterpiece

The rhetoric of Newton's first paper—rivetting to this day—was that of youth: brash and brilliant, relying for its persuasive effect on a clash of principles, a decisive confrontation, clearly and unequivocally presented. In contrast to his early papers, the rhetoric of the *Opticks* was that of late middle age: a canny and successful attempt to transform a youthful invention into a durable inheritance. In the *Opticks*, in the exposition of his theory, Newton employed a Euclidian arrangement to create an impression of historical continuity and logical inevitability. In addition, by piling experiment on experiment and, in each experiment, detail on detail, he created in this work an overwhelming presence for his experimental method. Finally, in the book's last section, he initiated a cascade of rhetorical questions whose cumulative effect was both to sanction his science and license his speculations. "Let me suggest," says historian of science David C. Lindberg,

> that optics from Alhazen to Newton (and, in some respects, from Aristotle and Ptolemy to Newton) must be seen as a continuously unfolding discipline. Historians of science have become overly conditioned to perceive the sixteenth and seventeenth centuries as a revolutionary period, during which scientists made a clean break with the past. There is, of course, much truth in this view, even in optics; but if pushed too far, it obscures the large measure of continuity in the history of optics from Alhazen onward. (1968, 36)

Lindberg's evolutionary interpretation of optical history has a Newtonian parallel. In February of 1676, in a letter to Hooke, Newton explicitly acknowledges his place in a long optical and scientific tradition: "What Des-Cartes did [the discovery of the sine law] was a good step. . . . If I have seen further it is by standing on [th]e sholders of Giants" (1959, 1:416).

There is a vast difference in intent between the historian and the scientist. Lindberg presents the essential continuity of optics as a historical interpretation, an end; but for Newton, in this early letter, historical continuity is a means, the outline of a rhetorical strategy. This strategy—so different from that of his early paper—is fully embodied nearly three decades later in his *Opticks*. Newton adopts Descartes's *topos*: his is an evolution, not a revolution.

In his early paper, the dominant order is narrative; in his *Opticks*, virtually the same material is set out as a Euclidian deduction. In the early paper, Newton uses narrative to dramatize the clash between past and present; in the *Opticks*, he uses Euclid to display the present as a deductive consequence of the past: "I have now given in Axioms and their Explications the sum of what hath hitherto been treated of in Opticks. For what hath been generally agreed on I content my self to assume under the notion of Principles, in order to [accomplish] what I have farther to write" (1979, 19–20).

In the *Opticks*, however, there is a tension between the deductive form Newton invariably adopts and the inductive order of knowledge he routinely employs. In the crucial first book, Euclidian form is particularly strict: definitions and axioms are followed by a series of propositions with their proofs. However, the method of all but the last proposition is analysis, induction rather than deduction; it "consists in making Experiments and Observations, and in drawing general Conclusions from them by Induction, and admitting no Objections against the Conclusions, but such as are taken from Experiments, or other certain Truths" (1979, 404). Only in the last proposition does the method change to synthesis or "Composition": the nature of light and of both radiant and natural color having been proved "by Experiments," they are now demonstrated by deduction (1979, 405, 20).

That this blurring is deliberate is clear from Newton's suppression of the explicit epistemological introduction to the first edition (Westfall 1984, 640–44). To move from the early papers to the *Opticks* is not to move from one science to another but from one rhetoric to another; to pass from a work that continually clarifies the priority of experiment in the order of knowledge to one that obscures the distinction between the traditional view of experiment as confirmatory and Newton's more radical view. For the initiated, the emphasis on experiment underscores its radical epistemological role, but this emphasis can bear a more conservative interpretation. It allows the French Cartesian Nicolas Malebranche to say that Newton's theory "fits . . . all his experiments" (Guerlac 1981, 110). In *Opticks*, Newton deliberately includes in his scientific audience even those who might differ from him on epistemological grounds.

The epistemological ambiguity of the experiments described in *Opticks* contrasts vividly with their clear and complete presentation. Throughout, the presence of Newton's experimental method is enhanced by the sheer number described and by the quantitative meticulousness with which their methods and results are reported. For any who might find experiments, for any reason, persuasive, Newton describes not two, as in his first paper, but dozens, all with a meticulousness arguably new to science: "I have set down such Circumstances, by which either the Phaenomenon might be render'd more conspicuous, or a Novice might more easily try them" (1979, 25). In his first paper, the *Experimentum Crucis* was meant to have its effect—an effect nothing less than the overturning of traditional optics—without the benefit of either a diagram or sufficient detail for easy replication. In contrast, in *Opticks*, a typical experiment is accompanied by a detailed diagram and a description that begins thus: "In the middle of two thin Boards I made round holes a third part of an Inch in diameter, and in the Window-shut a much broader hole being made to let into my darkned Chamber a large Beam of the Sun's Light" (1979, 45).

This concern for the rhetorical value of meticulous detail extends to measurement. Everywhere in *Opticks*, measurement, so important in the early papers, increases in importance; in seemingly every case, measurements previously made are remade: in book 2, part 2 of *Opticks*, for example, Newton recalculates the values of a table he published in *Transactions* nearly thirty years before (1978, 219; 1979, 233). The differences are sometimes very small indeed. In the earlier work, the thickness of glass at which indigo of the second order is most intense is .0000085; in *Opticks*, the figure is .000008182, a difference of 318 *billionths* of an inch!

There is one final rhetorical effect. Halfway through the first part of the third book, a remarkable communicative event occurs: both the Euclidian structure and the experimental program break off completely, and a long section of queries is initiated, each in the form of a negative rhetorical question. The question marks that punctuate these queries bestow upon Newton a double benefit: the question marks underline their speculative nature but leave undiminished their rhetorical force. By their very nature, negative rhetorical questions are strong positive assertions. In the body of *Opticks*, a light ray is defined as "the least Light or part of Light, which may be stopp'd alone without the rest of the Light, or propagated alone, or do or suffer any thing alone, which the rest of the Light doth not or suffers not" (1979, 2). The definition is operational: light behaves as if it was composed of rays made up of parts. But in the queries, light rays are unequivocally physical entities: "Are not the Rays of Light very small Bodies emitted from shining Substances?" (1979, 370).

The dual character of the queries permits Newton to meet some crucial objections of his early critics. Hooke and Pardies had criticized Newton's inference that light was a substance. In *Opticks*, Newton blunts this criticism by a distinction. In the body of this work, he confirms by experiment that light behaves consistently as if it was a substance; in his queries, he makes a plausible but not a scientific case that light is, in fact, a substance. Newton's distinction between science and speculation also blunts Huygens's complaint that Newton's explanation of light was incomplete because it did not include a specific mechanism for its propagation. In his *Opticks*, Newton shows his willingness to speculate about mechanisms so long as speculations are not confused with science. In the queries, for example, Newton considers the anomalous refractive behavior of light when passed through island crystal (a crystalline mineral now known as Iceland spar). From this behavior, he infers a physical property of rays of light, one that might account for the anomaly: "every Ray may be consider'd as having four Sides or Quarters, two of which opposite to one another incline the Ray to be refracted after the unusual manner" (1979, 360). However, this mechanical explanation is speculative, not scientific.

In his "Account," Newton says:

> In this Philosophy Hypotheses have no place, unless as Conjectures or Questions proposed to be examined by Experiments. For this Reason Mr. *Newton* in his Optiques distinguished those things which were made certain by Experiments from those things which remained uncertain, and which he therefore proposed in the End of his Optiques in the Form of Queries. (1980, 312)

Despite their speculative character, the queries serve certainty by staking out its limits. By clearly demarcating the opposing edges of science and speculation, the queries affirm the scientific status of the conclusions that precede them and the methods by which these conclusions were reached. At the same time, the rhetorical forcefulness of the queries, enhanced by the aging Newton's overwhelming reputation, clearly intend a constraint on the future of optical research. Newton sets his queries down so that "a farther search" may "be made by others" (1979, 339). In the body of *Opticks*, the properties of light are experimentally demonstrated and deductively proved; in the queries, the ground plan for the next century of optical research is successfully laid out.

Newton's rhetorical masterpiece became the model for optics and for experimental science in the next century. Initially, scientists did not have to believe in the epistemological priority of experiment or in the particle nature of light to accept Newton's empirical results and to acknowledge his scientific leadership. That acceptance and that acknowledgment eventually transformed optics into a modern experimental science along Newtonian lines; eventually, to do optics was to accept particle theory and to presuppose the priority of experiment in the order of knowledge.

A Complete Rhetorical Description of Scientific Change?

It is now time to deal with the objection broached in the introduction to this chapter that the transitions between Cartesian and Newtonian optics and between the optics of Newton's first paper and his final masterpiece depend on matters of fact outside the realm of rhetoric. Because the striking contrast between Descartes's circular and Newton's oblong spectrum must survive every paraphrase, it may be argued, it cannot be merely a matter of language to those who seriously assert it; it must be a claim about how the world is. Similarly, no legitimate paraphrase can fail to embody the conceptual equivalence between Newton's early paper and his *Opticks*, an equivalence that survives the complete transformation in his rhetoric. The scientific content of a text, it may be contended, is precisely that which survives in its legitimate paraphrases; there really is a scientific core even if it can never actually be freed from all linguistic embodiments.

This argument fails. Descartes's and Newton's optics are incompatible only because prior to interpretation, we know that no paraphrase of either is legitimate that omits their differing views of the shape of the spectrum; we can say Newton's optical papers and his *Opticks* share the same content only because prior to interpretation, we know that no paraphrase is legitimate that omits the oblong shape of Newton's spectrum. Paraphrase is not innate, not an aspect of linguistic competence; it is an interpretive achievement. We may be inclined to say that the scientific content of texts can be paraphrased because they contain a core of meaning, but just the reverse is the case: the core of meaning is what we paraphrase consistently. We paraphrase consistently because we have been persuaded so to paraphrase; conceptual equivalence or its absence is the product of rhetorical exchange within the relevant community of scientists.

There is no more dramatic evidence that the rhetorical negotiation of conceptual equivalence is the basis of paraphrase than the fiery disputes over priority that continually arise in the history of modern science. In a letter to the French chemist Charles Adolph Wurz, for example, August Kekulé argues that neither Wurz nor the English chemists William Odling and Alexander Williamson made a discovery that Kekulé attributes to himself, a result that was an important step toward the understanding of organic compounds; in other words, Kekulé asserts, his texts and those of Wurz, Odling, and Williamson are not mutually paraphrasable; they do not say the same scientific thing:

> Your lovely investigations on the glycols are even more recent; so they can have served just as little as a point of departure for my views, and indeed I only remarked that the *further development* of my views only became possible through your experiments. As for the views of Williamson and Odling in the papers you cited, I gladly acknowledge, and did so in my publication, that the ingenious considerations of these chemists formed the point of departure for my views. But whether the idea of the basicity of atoms *as we presently conceive it* is expressed "clearly" in these papers is less certain. In fact it doesn't seem so to me; I think that if one goes back that far, then no doubt Laurent and Gerhardt could be considered to be the originators of this idea. (Rocke 1993, 174–75; emphasis in original)

In Kekulé's formulation, Wurz's work depends directly on his, and Odling and Williamson are his precursors, not his rivals.

The negotiation of conceptual equivalence is also evident in a dispute over the discovery of the brain hormone TRF (thyrotropic-releasing factor). In this dispute, Roger Guillemin claimed priority, despite the thirty-seven-day lag between his results and those of his rival Andrew Schally. He based his claim on the deficiency of Schally's method. If this method was unreliable, as Guillemin claimed, then Schally's result was unreliable as well. Despite the outward likeness of their claims, they did not say the same scientific thing; one paper was not a paraphrase of the other (Latour and Woolgar 1986, 146–47; Wade 1981, 168–82). Guillemin expresses the difference to a *Scientific American* audience: "At length the complete structure of the natural TRF was obtained by high resolution mass spectrometry. It turned out to be the structure pGlu-His-Pro-NH_2. . . . TRF from pig brains was *subsequently* shown to have the same molecular structure as TRF from sheep brains" (32; my emphasis). In accord with these views, Guillemin cites only one article by Schally, a report that *followed* his. According to Guillemin, Schally's was not a discovery but a confirmation of another's discovery.

The Limits of the Rhetorical Analysis of Science

This chapter along with the previous one on taxonomy cannot hope to demonstrate unequivocally that a full rhetorical description of scientific texts is always available. These chapters do show that some commonplace—and some far from commonplace—problems that might undermine such an attempt have plausible solutions. None of this suggests that science is *only* rhetoric; no sane person could reach so bizarre a conclusion. These chapters can show no more than that, like sociology, history, and philosophy, rhetoric is a discipline; moreover, its disciplinary status entitles us to speak of all the written and visual records of the sciences from a rhetorical perspective.

Part Three

The Range of the Rhetorical Analysis of Science

6 The Generation of Scientific Knowledge

The Evolution of Evolution in Darwin's *Notebooks*

In the last two chapters, I discussed the extent and limits of rhetorical analysis; now I would like to turn my attention to its range, to rhetoric's ability to analyze significant cases of the generation of scientific knowledge, its initial certification as such by communities of scientists, its spread through these communities, and, finally, its incorporation into their practices. In this chapter, I turn attention to the generation of a central concept in modern biology—evolutionary theory. For the elucidation of its generation, I turn to Charles Darwin's *Notebooks*, perhaps the most detailed record we have of the birth of an important scientific idea.

On October 2, 1836, after its circumnavigation of fifty-seven months, the *Beagle* anchored off the English coast, and Darwin disembarked. In his kit was a small book bound in red leather, the *Red Notebook*, two-thirds full of geological notes. Busy disposing of the specimens he had gathered on the voyage and reacquainting himself with his family, Darwin did not immediately focus his attention on writing and thinking. But by the beginning of 1837, he was simultaneously converting his diary of the voyage into a travel book and making additional entries in the *Red Notebook*. These entries, these "brilliant scraps" made in the first half of 1837 (Herbert 1974, 247), contain Darwin's first speculations on evolutionary theory and led directly to his decision to begin the first of his transmutation notebooks.

The propriety of a rhetorical analysis of Darwin's *Notebooks* may be open to question: what warrant can there be for applying the classical categories of style and the order of arguments to these jottings, meant apparently for no audience, haphazardly arranged, and so abbreviated as to be, in many cases, barely comprehensible? The warrant rests on a theory: that our professional selves are the products of inner deliberation. This view is at one with our everyday metaphors of mentation: in the latter half of the *Red Notebook*, Darwin, we might say, was of two minds concerning evolution; in a series of entries, he debated with himself over its possibility. Charles Sanders Peirce turns these metaphors into philosophy when he says that a person's

> thoughts are what he is "saying to himself," that is, saying to that other self that is just coming into life in the flow of time. When one reasons, it is that critical self that one is trying to persuade . . . [Moreover,] man's circle of society (however widely or narrowly this phrase may be understood), is a sort of loosely compacted person, in some respects of higher rank than the person of an individual organism. (1955, 258)

Biographers, historians, philosophers of science, cognitive psychologists all have exhibited a legitimate interest in Darwin's notebooks as evidence of his intellectual development. But none has consistently treated these entries as a special genre: not public documents or letters meant for identifiable audiences; not diaries framed for the particular audience of one's future self; not first drafts of such documents, early attempts at audience accommodation; not Rorschach or TAT protocols, self-presentations to an audience of psychologists. The audience of the notebooks is Darwin, immediately, and Darwin only; their assertions and accompanying visuals are communicative acts of a special sort, acts whose status as knowledge is subject to marked epistemological fluctuation. As we move closer to the *Origin*, we experience a shift from verbal and visual embodiments instrumental to self-persuasion to those instrumental, potentially, in the persuasion of others. However, it is not in these later statements that the unique character of the notebooks lies; it is rather in the mental processes initiated in the most disjointed, least comprehensible entries and in the most primitive visuals. These most clearly display the special flavor of the notebooks: a disciplined lack of commitment to the full truth of assertions and their visual counterparts, a deficiency that enables the transformations to final theory.

This chapter traces the development of Darwin's ideas from these initial epistemologically fluctuating acts of self-persuasion to their final form—the pubic acts of persuasion embodied in the prose and in the single comprehensive visual that constitute the *Origin*.

Style as an Index of Self-Persuasion

There are two styles in the *Red Notebook*: the cryptic and the telegraphic. Although the telegraphic style maintains most of the norms of Victorian prose, it frequently omits articles, copulas, and the existential *there*; and it regularly employs standard abbreviations. (In the transcribed passages, I have marked Darwin's deletions "< >," his additions { }," editorial additions as "[ ]," and end of entry or page as "—"):

> Volcanos only *burst* out where strata in act of dislocation (NB. dislocation connected with fluidity of rock . . {in earliest stage} when covered up beneath ocean).—The first dislocations & eruptions can only happen during first movements, and therefore beneath ocean, for subsequently there is a coating of solidifying igneous rocks which would be too thick to be penetrated by the repeted trifeling injections. (1987a, 68; emphasis in original)

In addition to the telegraphic features, the cryptic style displays an absence of connectives, especially logical connectives; deviations from syntax radical enough to render semantic connections problematic; and radically deviant punctuation:

> Speculate on neutral ground of 2 ostriches; bigger one encroaches on smaller.—change not progressif<e>: produced at one blow. if one species altered: <altered> Mem[ento] [Remember]: my idea of Volc: islands. elevated. then peculiar plants created. if for such mere points; then any mountain. one is falsely less surprised at new creation for large.—Australias = if for volcanic isl[d.] then for any spot of land. = Yet new creation affected by Halo of neighbouring continent: ≠ as if any | creation {taking place} over certain area must have peculiar character. (1987a, 61)

In this passage, Darwin seems to be speculating on the possibility that evolution is saltational. Although Darwin, for whom alone this passage is written, undoubtedly could interpret it without ambiguity, my interpretation may easily be questioned. While the telegraphic style is a shorthand, designed like all shorthands to close the gap between the rate of verbal production and its transcription, the cryptic style has more complex roots. Built up chunk by linguistic chunk, each framed by terminal or near-terminal punctuation, passages in this style actively resist interpretation. It is difficult to see how these linguistic strings could have issued from the operation of syntactic rules on semantic components stored simultaneously in short-term memory. More plausibly, they represent a stage prior to the application of these rules. Under this construal, each chunk must have been retrieved individually with effort and must have been simultaneously *thought* and *uttered.*

Consider these two utterances separated by less than a hundred words in which Darwin refers to his discovery of a petrified forest in the Andes as unmistakable evidence that this mountain fastness had once been a coastline:

> Mem[ento]. SUBSIDENCE Uspallata of which no trace except by trees
>
> Let it not be overlooked that except by trees, I could not see trace of Subsidence at Uspallata. (1987a, 69–70; see also 1987b, 406–7)

According to my reading, on the self-command—"Mem[ento]" or "Remember"—the four linguistic chunks in the first utterance are called forth sequentially from long-term storage, the second by the first, the third by the second, the fourth by the third. On this interpretation, to read the *Red Notebook* is to experience the psychological reality of a generative grammar: the second linguistic string, the sentence, is generated from the first by the application of the syntactic rules of English to material that hovers between language and pure thought.

In the *Red Notebook*, we reach an important limit beneath which rhetorical analysis cannot operate. Entries in the cryptic style approach a threshold below which interpretation must give way to causal analysis: language much more disordered than this would cease to fulfill the norm of comprehensibility that is communication's essential condition. A style more disordered than the cryptic would also be excluded from rhetorical theory because

> someone cannot have a belief unless he understands the possibility of being mistaken, and this requires grasping the contrast between truth and error—true belief and false belief. But this contrast . . . can emerge only in the context of interpretation, which alone forces us to the idea of an objective, public truth. (Davidson 1984, 170; see also Quine 1976, 235)

In the *Red Notebook*, as Darwin's speculations approach the stability of his final formulations, his style approximates the norms of his published prose. As his style in these speculations approaches these norms, his utterances simultaneously approach the norms that speech-act theory stipulates for assertions in the public domain: that they be clear, that they be true, and that we take responsibility for their truth. The epistemic status of entries in the *Red Notebook* varies: the nearer these speech acts approach to the cryptic style, the closer they approximate the suspension of the validity claims common to public assertions. This suspension indicates that, early in the development of his ideas, Darwin is prepared to fix neither reference nor belief. Neither confusion nor irresponsibility is at work; on the contrary, only so long as he remains uncommitted to the full truth of his assertions can Darwin maintain the prolonged suspension of belief essential to the acceleration of internal conceptual change. In Darwin's private mental space, style abandons its public function; it is no longer a means of persuasion. Because it provides evidence for the status of entries as knowledge, style becomes, instead, an index of the relative stability of the beliefs Darwin espouses.

Order as Index of Intellectual Development

The order of entries in the *Red Notebook* is also an index of the progress of Darwin's thought: their arrangement anticipates his intellectual development. Themes later linked in important theoretical formulations appear separately, form sequences, and become intertwined. In the *Red Notebook* entries made soon after his return, Darwin focuses on a range of topics that were to form two major themes of his intellectual career: the causes of geological change and the biogeographical distribution of species in space and time. The following outline represents my reconstruction of these notebook themes.

I. THE CAUSES OF GEOLOGICAL CHANGE

- A. Igneous change
    - 1. Subsidence and elevation of the earth's crust
        - a) The creation of mountains, volcanoes, and earthquakes
        - b) Their pattern
        - c) The action of volcanoes and earthquakes
    - 2. The action of lava
- B. Aqueous change
    - 1. Silting to build up land
    - 2. Denudation to wear down land
    - 3. Transportal of boulders
    - 4. Glacial action

II. THE BIOGEOGRAPHICAL DISTRIBUTION OF SPECIES IN SPACE AND TIME

- A. Distribution of flora and fauna in space
    - 1. Effect of climate
    - 2. Effect of barriers
    - 3. Question of adaptation
- B. Distribution of flora and fauna in time
    - 1. Extinction of species (fossil record)
    - 2. Creation of species: descent

Although the entries in the *Red Notebook* are far less organized than this outline suggests, their order, nevertheless, exhibits trends that anticipate Darwin's later, mature formulations in *Origin*. The list below is a representative sequence of thirty-one entries from the *Red Notebook*; the entries are coded according to the above outline and are arranged into fourteen groups.

1. IA, IA
2. IIA
3. IA, IA
4. IIA
5. IIB, IIB, IIB, IIB
6. IIA
7. IIB
8. IIA, IIA, IIA, IIA, IIA, IIA
9. IIB, IIB
10. IA
11. IIB, IIB
12. IIA, IIA
13. IB
14. IA, IA, IA, IA, IA

In this display, two contrasting tendencies are apparent. In the first, Darwin intersperses the geological and the biogeographic; in the second, he pursues the same theme in a single run of entries. The tendency to intersperse anticipates the close connection between geology and biogeography in Darwin's mature theory; the tendency to separate anticipates Darwin's decision to bifurcate his intellectual efforts: after filling the *Red Notebook*, he simultaneously began *Notebook A* on geology and *Notebook B* on the transmutation of species.

There is a third tendency no outline can represent: short, isolated entries on particular themes are followed by longer passages in which these themes appear in concert. Witness the first two biogeographical entries in the run IIA (group 8 in the above list), both on the distribution of flora and fauna in space:

> 1. Webster Antarctic veg:-
> 2. no. mad dogs. Azores. although kept in numbers.
>
> (1987a, 60)

These are followed by a longer biogeographical entry concerning the distribution of flora and fauna in time, one in which Darwin speculates that a worldwide drought may have led to the common burial of species with normally divergent territories:

> Consult W. Parish. & Azara about dry season[.] 1791. seen commonly bad over whole world. {Was this so in Sydney, consult history? Phillips} | 1826.27.28 grt. drought at Sydney. which caused Capt. Sturt expedition.—| ¿ Another one in 1816(?).—

Four pages later, Darwin links species distribution in space and time in a powerful single formulation:

> The same kind of relation that common ostrich bears to (Petisse. & diff kinds of Fourmillier): extinct Guanaco to recent: in former case position, in latter time. (or changes consequent on lapse) being the relation. As in first cases distinct species inosculate [separate], so must we believe ancient ones { ∴ } not *gradual* change or degeneration. from circumstances: if one species does change into another it must be per saltum [by leaps]—or species may perish. = This <inosculation> {representation} of species important, each its own limit & represented.—Chiloe creeper: Furnarius. <Caracara> Calandria: inosculation alone shows not gradation: — (1987a, 62–63)

In this passage, we have moved past thematic connections into a relationship of natural necessity, a theory about the natural world. This traversal from theme to theory is tellingly exemplified by the {∴}, the "therefore" that Darwin added as an afterthought, a {∴} that marks the sudden inception of his awareness of a necessary relationship, a natural law.

As the passages in which themes intertwine approach the stylistic and organizational norms of Darwin's finished prose, their theoretical content verges on the norms of his mature theory: the distribution of flora and fauna in space and time subject to the same natural law. When we examine the arrangement of the *Red Notebook*, then, we see themes become theory, but in the passages in which themes previously separate are closely linked, we leave the province of arrangement proper and pass over into argument, turning fact into theory.

Turning Fact into Theory

Around August 1833, Darwin found himself in Northern Patagonia. While there, he heard of a rare form of ostrich, the Avestruz Petise (subsequently identified as a rhea), a bird smaller than the common ostrich, darker in color, and shorter of leg. Around Christmas of the same year, he sat down to a dinner of ostrich shot by Conrad Martens, the ship's artist. At first, Darwin thought the bird was a young specimen of the common sort of ostrich; after the meal, however, his memory of the Petise returned, and he saved the head, legs, and several feathers of the bird he and his shipmates had just consumed. In notes probably written within three months of the incident, Darwin recounts his change of mind. He points out the differing range of the smaller bird: "Whatever Naturalists may say, I shall be convinced from such testimony. as Indians & Gauchos, that there are two species of Rhea in S. America" (1963, 274).

At the time, Darwin had collected the specimen as part of his routine duties as the expedition's naturalist. By March of 1834, however, he had singled out the species for special consideration: "But what is of more general interest is the unquestionable (as it appears to me) existence of another species of ostrich, besides the Struthio Rhea. —" (1985—, 1:370; see also 1977, 8). Two years later, in mid-1836, on the last leg of his journey home, Darwin mentions the Petise again in a series of ornithological notes: "In conclusion, I may repeat that the Struthio Rhea inhabits the country of La Plata as far as a little south of the R. Negro, in Lat. 41°: & that the Petise takes its place in Southern Patagonia, the part about the R. Negro being neutral territory" (1963, 272; Sulloway 1982, 337).

On March 12, 1837, in his first year back in England, Darwin reports that ornithologist John Gould has confirmed his identification, naming the new species in honor of its discoverer (1985—, 2:11). Two days later, Gould reported his identification to the Zoological Society of London; at the same meeting, Darwin read a paper on the subject. He avoided speculation, stating, according to the secretary, "that the *Rhea Americana* inhabits the country of La Plata as far as a little south of the Rio Negro, in lat. 41°, and that the *Petise* takes its place in Southern Patagonia" (1977, 40). At this point, the newly named *Rhea darwinii* disappears from Darwin's public writing.

But not from his private writing. In the days following his presentation to the zoological society, Darwin initiated in private the evolutionary speculation he eschewed in public: he recorded in the *Red Notebook* an entry singling out the neutral ground between the ranges of the two species as a site for theorizing (Sulloway 1982, 381): "Speculate on neutral ground of 2 ostriches." That two species of ostrich shared contiguous territory was a fact destined to have a profound impact on Darwin's thought. In his *Autobiography*, he says of his most creative period: "It was evident that such facts as these . . . could be explained on the supposition that species gradually become modified; and the subject haunted me" (1958, 118–19).

Within the next two months comes the first product of this speculation. It concerns what is to become a recurring theme, descent with modification: "When we see Avestruz two [ostrich] species. certainly different. . . . Yet one is urged to look to common parent? why should two of the most closely allied species occur in same country?" A slightly later entry contains a far more assertive reference to descent with modification: "I look at two ostriches as strong argument of possibility of such change, —as we see them in space, so might they in time" (1987a, 70, 175). The question marks are gone, the syntax is regular though abbreviated, symptoms that private persuasion has been completed, that a belief has become a relatively stable component of the network of theoretical statements that constitutes Darwin's professional self.

There is a second theme of Darwin's speculation: the importance of "neutral ground" in the determination of good species. Around mid-May 1838, the ostriches become an example of a clearly formulated rule for species determination:

"analogy from every country & class tells us that ¿O[petiorhynichus]. Modulator & O. Patagonicus. till neutral ground ascertained, call them varieties. but two ostriches good species because interlock [without interbreeding]." An entry written a bit later suggests a necessary condition for such speciation: "ISOLATION of range . . . tends to alteration views.—ostriches d[itt]o—" (1987a, 237, 277; 304).

By mid-1838, then, only months before the Malthusian-influenced revelation concerning natural selection as the cause of speciation, Darwin's speculations concerning the neutral ground shared by two ostriches have stabilized: variations produced by descent with modification and isolated by geographical barriers may form good species. These will not interbreed, even if, at some later time, their ranges become contiguous.

Only after a hiatus of twenty-two years in the *Origin of Species* are the ostriches presented again to the public. But in the *Origin* they are introduced, not as the source of important theoretical conclusions but as illustrations of the geographic succession of closely allied species, a phenomenon that can be best accounted for by descent with modification:

> the naturalist in travelling . . . from north to south never fails to be struck by the manner in which succesive groups of beings, specifically distinct, yet clearly related, replace each other. He hears from closely allied, yet distinct kinds of birds, notes nearly similar, and sees their nests similarly constructed, but not quite alike, with eggs coloured in nearly the same manner. The plains near the Straits of Magellan are inhabited by one species of Rhea (American ostrich), and northward the plains of La Plata by another species of the same genus; and not by a true ostrich or emeu, like those found in Africa and Australia under the same latitude. . . . We see in these facts some deep organic bond, prevailing throughout space and time, over the same areas of land and water, and independent of their physical conditions. The naturalist must feel little curiosity, who is not led to inquire what this bond is.
>
> This bond, on my theory, is simply inheritance. (1964, 349–50)

The significant distinction in argument between the *Notebooks* and the *Origin* lies in the contrast between what convinced Darwin and what Darwin thought would persuade his first audience. This is the only explanation necessary for a range of differences between the notebooks and the *Origin* in the way arguments are built. In the succession of notebook entries on the rhea, for example, is enacted a rhetorical drama—self-persuasion; in the shift from the notebooks to the *Origin*, we see the difference between self-persuasion and the persuasion of others—a rhetorical difference. This contrast applies also to the most striking difference in argumentative strategy between the notebooks and the *Origin*: that the analogy between artificial and natural selection, which looms so large in the latter, is of such trifling importance in the former.

From Theory to Theory

In this section, I focus on the development of Darwin's theory of evolution as it is embodied in a series of tree diagrams that begins with *Notebook B* and ends in the *Origin*. This series raises the question of whether this intellectual development can be viewed as conceptual evolution in the strict sense, the differential survival of an evolutionary theory in the face of selection pressures from a changing conceptual environment. If this question can be answered in the affirmative, the series will also depict a rhetorical evolution parallel to that in the previous section: as we move from the diagrams in the notebooks to the single diagram in the *Origin*, we move from diagrams integral to self-persuasion to those designed for the persuasion of others.

In "Picturing Weismannism: A Case Study of Conceptual Evolution," James R. Griesemer and William C. Wimsatt apply evolutionary theory to the facts of change in a particular case, a set of diagrams that represent August Friedrich Leopold Weismann's central doctrine of the continuity of the germ line. In Griesemer and Wimsatt's view, scientific concepts are not the best unit for tracing conceptual development. Because concepts are expressed in a natural language, they are hard to pin down; moreover, once pinned down, their meaning is still often subject to extended debate. In addition, because they are expressed in words, concepts present no clear boundary between themselves and the surrounding text. As a result, the analysis of the relationship between them and their "environment" is difficult if not impossible to determine. These defects also make concepts hard to compare in detail or to array in lineages. Griesemer and Wimsatt argue that because diagrams suffer from few of these defects, they are an ideal "organism" with which to demonstrate conceptual evolution (1989, 16–21). Weismann diagrams, they contend, behave in good Weismannian fashion: "Theory-phenotypes do not directly make more theory phenotypes in the scientists which they infect. Rather what is passed on is a relatively compact generative structure which, in the right conceptual and social environment, generates the theory-phenotype of the next generation" (1989, 14). In other words, descent does not necessarily yield exact copies but a variety of offspring with the same generative structure.

Following Griesemer and Wimsatt, I contend that the history of Darwin's tree diagrams is an evolutionary history. Between May and July 1837 in *Notebook B*, Darwin used pictures and words in pursuit of a theory that could account for descent, divergence, and extinction (his later addition is enclosed in double brackets). In this diagram (see fig. 6.1) (1987a, 177), the dotted lines of the trunk and the origin of the branches represent the death of earlier species as a consequence of the evolutionary process.

The tree of life should perhaps be called the coral of life, base of branches dead; so that passages cannot be seen.— this again offers

contradiction to constant succession of germs in progress.— «no only makes it excessively complicated.»

Fig. 6.1. Darwin's Earliest Tree from *Notebook B.* Reprinted from Charles Darwin: *Charles Darwin's Notebooks, 1836–1844: Geology, Transmutation of Species, Metaphysical Enquiries,* edited by Paul H. Barrett, Peter J. Gautrey, Sandra Herbert, David Kohn, Sydney Smith. Copyright © 1987 by Paul H. Barrett, Peter J. Gautrey, Sandra Herbert, David Kohn, Sydney Smith. Used by permission of the publisher, Cornell University Press.

Is it thus fish can be traced right down to simple organization.—

birds— not.

Fig. 6.2. Darwin's Second Tree from *Notebook B.* Reprinted from Charles Darwin: *Charles Darwin's Notebooks, 1836–1844: Geology, Transmutation of Species, Metaphysical Enquiries,* edited by Paul H. Barrett, Peter J. Gautrey, Sandra Herbert, David Kohn, Sydney Smith. Copyright © 1987 by Paul H. Barrett, Peter J. Gautrey, Sandra Herbert, David Kohn, Sydney Smith. Used by permission of the publisher, Cornell University Press.

Right after the diagram of figure 6.1, Darwin makes a second point, illustrated again by a visual (see fig. 6.2) (1987a, 177). In this diagram, the solid lines represent the evolutionary development for which we have evidence; the dotted lines, development for which we do not. While evidence for the evolutionary development of the fish is available, it is not available for birds. As a consequence, common origin in the latter case must remain a hypothesis.

Ten pages later, Darwin writes "I think" and produces not an assertion but a third version of his branching diagram (see fig. 6.3) (1987a, 180). There are no further diagrams of this sort in the notebooks. None is needed, as this last already includes the key concepts of descent, diversity, and extinction, with descent represented linearly, diversity by horizontal distance, and extinction by branches without a foot. In figure 6.3, the (1) represents an origin, and the A, B, C, and D, presumably, genera or species.

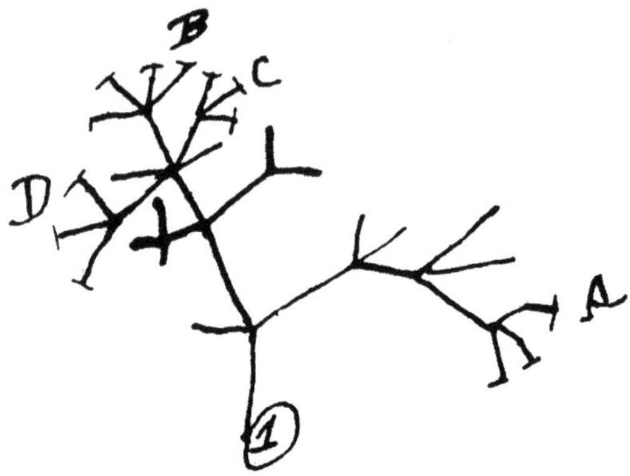

Fig. 6.3. Darwin's Third Tree from *Notebook B*. Reprinted from Charles Darwin: *Charles Darwin's Notebooks, 1836–1844: Geology, Transmutation of Species, Metaphysical Enquiries*, edited by Paul H. Barrett, Peter J. Gautrey, Sandra Herbert, David Kohn, Sydney Smith. Copyright © 1987 by Paul H. Barrett, Peter J. Gautrey, Sandra Herbert, David Kohn, Sydney Smith. Used by permission of the publisher, Cornell University Press.

Shifting attention from the notebooks to the *Origin*, the arena moves from one in which Darwin's concepts evolve to one in which his diagram itself evolves as an expression of his theory, a rhetorical device designed to enhance that theory's persuasive impact. This transformation begins in the four related diagrams (figs. 6.4, 6.5, 6.6, and 6.7) in Darwin's "big book" on evolutionary theory, "Natural Selection," essentially his first draft of the *Origin*. Each diagram represents a single aspect of the theory.

The first, Diagram I (fig. 6.4), is devoted to descent through natural selection. To create this diagram, the last figure (fig. 6.3) in *Notebook B* has been rotated 180 degrees; moreover, its "root," the (1) in figure 6.3, is now multiple, represented by letters A through M. Descent is represented linearly, diversity by horizontal distance, and extinction by the end of a line, as in B, K, and L (1975, 236).

Diagram II (fig. 6.5), illustrates the importance of natural selection in ensuring diversity; in the absence of natural selection, variation is severely restricted, a restriction illustrated in narrow columns of descent (1975, 237).

Diagrams III and IV (figs. 6.6 and 6.7, respectively) focus on special problems of the theory. Diagram III (fig. 6.6) concerns itself with evolutionary distance as a consequence of divergence: the divergence of A (a^{10}, h^{10}, l^{10}) after ten generations compares with the lack of divergence in M (M^{10}). Selection pressures have forced species A to diverge, while species M has been relatively free from such pressures, a freedom that works against evolutionary change (1975, 237).

Diagram IV (fig. 6.7) (1975, 237) contrasts extinction (B, C, D, G, H, I, K, L), divergence through evolutionary descent (A and M), and descent without diversity, "living fossils" (E, F).

Beginning with its first appearance in *Notebook B*, then, Darwin's diagram evolves from a simple depiction of branching to a full depiction of the essentials of evolutionary theory, a depiction that culminates in the single comprehensive diagram in the *Origin* (fig. 6.8).

In its first interpretation (fig. 6.8) in chapter 4, the diagram's vertical dimension is time; its horizontal dimension is organic diversity. The farther one gets toward the top of the drawing, the nearer one gets to the present; the farther apart species are, the more they have differentiated themselves from their parent species through evolutionary processes. Reading the diagram vertically, we trace descent. We see that a^1 and m^1 and a^{10} and m^{10} have the same remote ancestor A. Reading horizontally, we see that far greater evolutionary divergence exists between the latter pair than between the former.

Extinction is also represented in the diagram, literally, by the end of a line. Species s^2, a descendant of A, becomes extinct in era II; species u^8, a descendant of z^4, which is a descendant of I, becomes extinct in era VIII. The diagram also embodies the cause of extinction, the struggle for existence. It is no accident that A and I, whose ancestors are still living, were originally nearly the most divergent from each other, for the nearer species are along the horizontal dimension, the less "space" that exists between them, the more likely they are to compete for the same territories and resources in the battle for survival; the more likely they are to become extinct. As a general rule, species such as A, whose progeny survive, diverge. But survival does not always depend on divergence. B, C, D, E, F, G, H, K, and L failed to diverge and went extinct. But although F also failed to diverge, it maintained itself through the years and survives as a living fossil.

Finally, Darwin makes a theoretical point about the significance of the numbered letters: "In our diagram the line of succession is broken at regular intervals by small numbered letters marking the successive forms that have become sufficiently distinct to be recorded as varieties. But these breaks are imaginary, and might have been inserted anywhere, after intervals long enough to have allowed the accumulation of a considerable amount of divergent variation" (1964, 119). Despite their individuation as points, species a^1, a^2, a^3, etc., are not real;

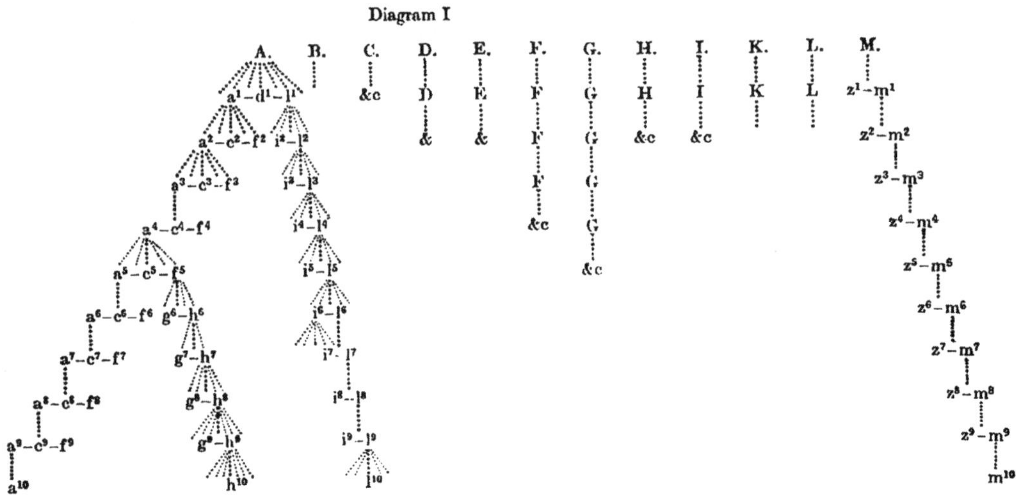

Fig. 6.4. First Tree from Darwin's "Big Book." Reprinted figure from *Charles Darwin's Natural Selection*, ed. R. C. Stauffer, Cambridge: Cambridge University Press, 1975, pp. 236–37, copyright © Cambridge University Press. Reprinted with the permission of Cambridge University Press.

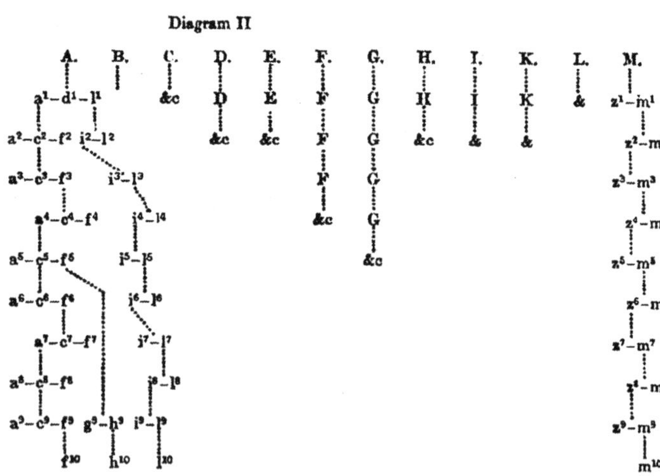

Fig. 6.5. Second Tree from Darwin's "Big Book." Reprinted figure from *Charles Darwin's Natural Selection*, ed. R. C. Stauffer, Cambridge: Cambridge University Press, 1975, pp. 236–37, copyright © Cambridge University Press. Reprinted with the permission of Cambridge University Press.

Original Species in Diag. I			A.	B.	C.	D.	E.	F.	G.	H.	I.	K.	L.	M.
Diagram III	a^{10}.	h^{10}.	l^{10}.	B.	C.	D.	E.	F.	G.	H.	I.	K.	L.	M^{10}

Fig. 6.6. Third Tree from Darwin's "Big Book." Reprinted figure from *Charles Darwin's Natural Selection*, ed. R. C. Stauffer, Cambridge: Cambridge University Press, 1975, pp. 236–37, copyright © Cambridge University Press. Reprinted with the permission of Cambridge University Press.

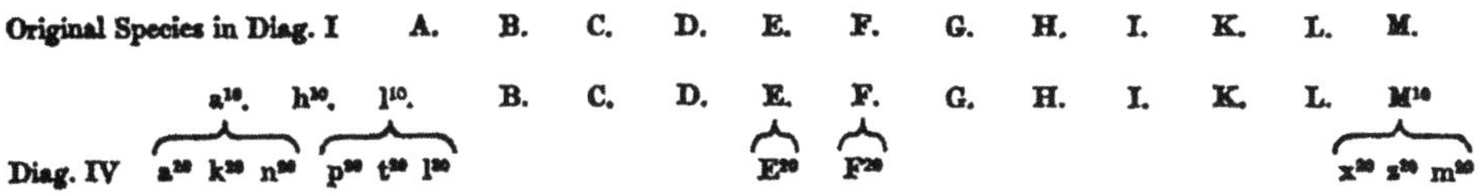

Fig. 6.7. Fourth Tree from Darwin's "Big Book." Reprinted figure from *Charles Darwin's Natural Selection*, ed. R. C. Stauffer, Cambridge: Cambridge University Press, 1975, pp. 236–37, copyright © Cambridge University Press. Reprinted with the permission of Cambridge University Press.

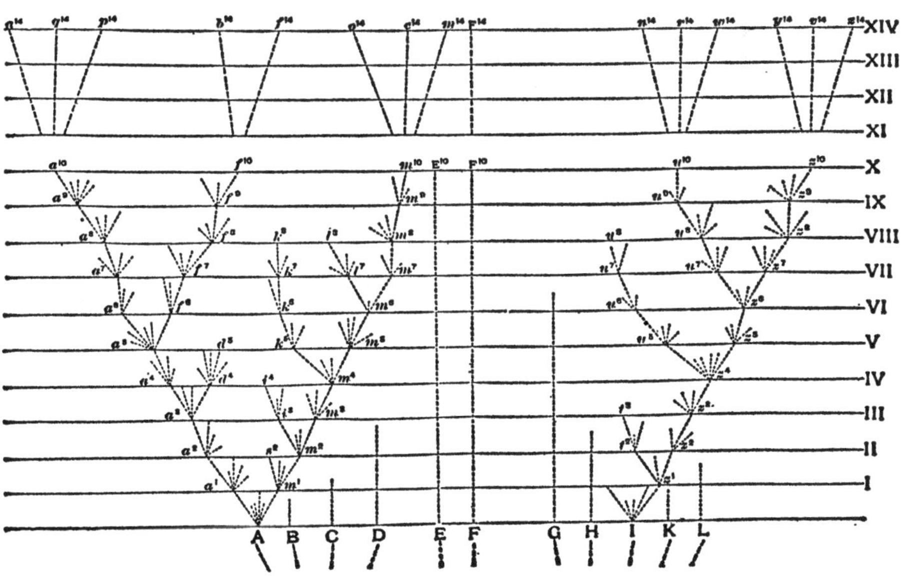

Fig. 6.8. The Tree from Darwin's *Origin.*

even the lines of divergence are conjectural; they are *dotted* lines. Only the processes that create divergence are real. Evolution embodies an ontology of process rather than product.

Moreover, as the time scale of the process is open, the amount of evolutionary change each incremental coordinate represents may legitimately vary. Darwin says: "If we suppose the amount of change between each horizontal line in our diagram to be excessively small, these three forms may still be only well-marked varieties . . . but we have only to suppose the steps in the process of modification to be more numerous or great in amount, to convert these three forms [a^{10}, f^{10}, and m^{10}] into well-defined species" (1964, 120). The time scale can also change, not incrementally, but by an order of magnitude. Darwin asks us to imagine that each horizontal line represents "a million or hundred million generations" (1964, 124). This variation in time scale and, consequently, in the span of evolutionary change blurs distinctions between species and variety, genus and species. In effect, Darwin turns the diagram into an argument for these intellectual leaps.

Thus far, we have seen Darwin's diagram as a representation of evolution, but it is equally a representation of the geological record, "a succession of the successive strata of the earth's crust including extinct remains" (1964, 124), a record of evolution. In the first case, the vertical dimension is metaphoric: space represents time. In the second case, the vertical dimension is denotative: space represents only space—the geological column. But the geological column is, as it were, time made visible; therefore, the vertical dimension also represents time. Really, it *is* time: looking at the fossil embeddings in the geological column, we are in a very real sense looking at the past. It is a past, moreover, that the first reading of the diagram allows us to understand. Looking at the geological record, Darwin says, "We find . . . such evidence of the slow and scarcely sensible mutation of specific forms, as we have a just right to expect to find" (1964, 336). The location of these forms and their fossil remains, it should be noted, is not their physical but their evolutionary location, not where they are found but where once found they belong in the evolutionary sequence.

Darwin's diagram combines geology and evolutionary theory into a rhetorical construction that blends two very different ancestors—the genealogical trees routinely displayed by royal and noble houses and contemporary scientific depictions of the geological column. To clarify the meaning of such blending inheritance, we may turn to an analogy with ordinary pictures: "What happens when two pictures are put together? Do they tend to overlap, or is one superimposed on the other, or do the features of both contribute to the product? Is each feature modified, or are various features of each picture maintained intact? . . . The blending inheritance of cultural replicators may sometimes result in the generation of novel variants outside of the existing range" (Heyes and Plotkin 1989, 159). In Darwin, both features—the tree and the column—contribute to the final product: we have a case of blending inheritance, the generation of a novel variant that incorporates the diagrams in the Big Book but goes beyond them, fusing genealogy and geology. In the process, the diagram has been divested of most of those features that facilitate pattern recognition in the case of middle-sized objects—

real trees and real geological columns. It is precisely this quasi-idealization that permits the diagram to mediate between the world and its explanation.

Darwin's is a combined rhetorical strategy. In his crucial fourth chapter of *Origin*, he does not rely solely on verbal argument or graphic presentation but avails himself of both. As a consequence of this dual rhetorical assault, the visual-textual compound that is the single diagram in Darwin's *Origin* becomes more than a depiction of evolutionary theory; it also becomes an argument in favor of that theory. The geological column is evidence for the theory of descent; the theory of descent explains the geological column. So interpreted, it may be said, Darwin's diagram *is* his theory, his argument. On this view, once we understand the diagram, we understand Darwin. On this construal, a view of the *Origin* as a book with one diagram is mistaken; it is really a diagram so complex that it takes a whole book to unfold its meaning. Such a view, however, misrepresents the division of labor that the fourth chapter so elegantly exemplifies. Necessarily, Darwin's diagram falls short of depicting the complexity of the processes involved. It altogether omits sexual selection, the effects of inbreeding that would retard change, and the environment that is change's motor. Nevertheless, its combination of conceptual complexity and graphic simplicity remains impressive.

The Origin of Species evolves in two phases: it begins with a rhetorical transaction within the self and ends in a rhetorical transaction with others. The notebooks enact a drama of self-persuasion: driven forward by the rush of new concepts and facts, held in check by the need to maintain the self as a coherent network of beliefs. As Darwin's theory evolves, his style alters from a means of expression in close touch with primary mental processes to one that anticipates public forums. Coincident with this stylistic process, the arrangement of the notebooks alters in character: isolated facts and concepts coalesce, intertwine, and, finally, unite into more and more fundamental law-like statements. Coincident with this, Darwin's visuals evolve to the point of clearly embodying the fundamentals of his theory. In the *Origin*, this theory operates in a different arena. In that arena, the *Origin*'s text and its single diagram represent a division of rhetorical labor: each medium of communication takes advantage of characteristics special to it, characteristics that converge in single persuasive construction.

7 The Initial Certification of Scientific Knowledge

Peer Review as the Institutionalization of Consensus

In the last chapter, I focused on scientific knowledge generated by an individual, Charles Darwin. But calling the work of Darwin scientific knowledge begs the question. This is because scientific knowledge is ineliminably social, that is, it is neither scientific nor knowledge until it is the conviction, not merely of the individual or even a small group of like-minded adherents but of a broad consensus of practitioners. Such a consensus rests on an even broader consensus already in place, that of entire human communities, as Donald Davidson makes clear. In his view, the three varieties of knowledge—of self, of other minds, and of the physical world—are interdependent; they stand or fall together:

> Communication, and the knowledge of other minds that it presupposes, is the basis of our conception of objectivity, our recognition of a distinction between true and false belief. There is no going outside this standard to check whether we have things right, any more than we can check whether the platinum-iridium standard kept at the International Bureau of Weights and Standards at Sèvres, France, weighs a kilogram. We can, of course, turn to a third party and a fourth to broaden and secure the interpersonal standard of the real, but this leads not to something intrinsically different, just more of the same, though the augmentation may be welcome. (1991, 164)

To know something is to know it communally, under a certain description, just as to affirm a proposition is to affirm the existence of the world of which the proposition is true, and to have affirmed, through the use of a language, the existence of other minds: "A community of minds is the basis of knowledge," Davidson says in summary; "it provides the measure of all things" (1991, 164).

To call something not just knowledge but scientific knowledge is to raise the bar, to say that consensus must be based on the rationally motivated agreement of a community of scientists. This agreement means that claims will be the product of agreed-upon procedures and that they will be supported by relevant argument if challenged.

Peer review, the subject of this chapter, is the institutionalization of this scientific consensus. Although in peer review, judgments that the standards of this scientific consensus are met may be made by only two or three scientists, this small number must not mislead us: despite their empirical status as fallible reflections of community will, these judgments are, politically speaking, representative. This does not mean that their fallibility is not apparent: the products of positive peer-review judgments are treated not as knowledge per se but as candidate knowledge. In this chapter, I scrutinize the process by which these judgments are reached, the consensus that creates candidate scientific knowledge.

There are explanations of peer review that do not insist on rationally motivated agreement. To system theorists, peer review is a self-regulating mechanism that protects the broad range of researchers from bad science and from more sweeping and speculative claims. This filtering prevents "the capricious redirection of goals, the proliferation of research programmes and the scattering of resources" (Myers 1985, 623). For system theorists, while system maintenance is real, rationality is an artifact of systems operation. An analysis according to systems theory is an analysis of the delicate balance between competing claims in the interest of system self-maintenance.

Game theory is another explanatory possibility for peer review. In this theory, negotiation is central, a procedure based not on rationally motivated agreement but on strategic maneuvering. When we apply game theory to peer review, we discover that the more general the claim, the more problematic the negotiation. As an illustration of this view, linguist Greg Myers recounts the publication histories of two biological papers. In each case, the scientist-author tries to obtain acceptance for the broadest possible claims in the most prestigious professional forums: "Both the biologists I am studying," Myers says, "try to make the highest level claim the editors and reviewers will allow" (1985, 602). For game theorists, power is the reality, and rationality an artifact of its operation. An analysis according to game theory is an analysis of strategies designed to persuade, to win the game.

Jürgen Habermas is critical of both systems theory and game theory because of their neglect of rationally motivated agreement. He does not deny that systems like peer review embody self-regulation nor that in peer review power and strategy are at play. From his point of view, however, systems theory is defective as an explanation because its self-regulating mechanisms can account for communicative action only by "objectivication," that is, by the reduction of such action to a systems function from which it differs in essence. To Habermas, we also err when we reduce communicative action to a contest for power; when we do so, we confuse an interaction whose goal is rationally motivated agreement with one whose goal is success at the price of such an agreement (1987, 280; see also 1984, 292).

For Habermas, to equate communicative action with the exercise of power is to disallow a crucial differentiation, the "strict distinction between *ties that are motivated empirically* through inducement and deterrence and *trust that is motivated rationally* through agreement based on reasons" (280; emphasis in original). To equate communicative action with system maintenance or gaming is to remain insensitive to real differences.

Habermas's preferred explanation of a phenomenon such as peer review is the *Ideal Speech Situation*, a theoretical formulation derived from speech act theory. The Ideal Speech Situation purports to be a set of presuppositions unavoidable if we are to create a rational consensus based on mutual understanding, to detect its presence, and to judge its quality. Perfect symmetry is required: interlocutors must reveal themselves, they must have every opportunity to initiate communication and to make assertions, and they must have equal control over the exchange; in the pursuit of rational consensus under these conditions, bias is revealed, countered, neutralized. From the point of view of the Ideal Speech Situation, although the knowledge that science produces is in constant flux and although at any one point, systems or game theory may be fully explanatory, over the long run, consensus is rationally motivated.

While the Ideal Speech Situation privileges consensus, it is also comfortable with the dissensus that often prevails. Because the Ideal Speech Situation defines rationality as a process, it is not embarrassed by seemingly wide differences, such as the diverse referee judgments that routinely characterize peer review. If rationality is a process, we are wedded not to the truth of particular claims but to communicative actions whose by-product is rationally motivated consensus, a consensus in science concerning our best accounts of the causal structure of the world.

The Ideal Speech Situation has been criticized because it ignores the realities of communicative action. Especially, the sufficiency of its presuppositions has been questioned because the theory omits any consideration of such characteristics of the participants as their intelligence or their competence; in addition, the Ideal Speech Situation omits the cultural traditions in which interlocutors participate and the relative economic and political power they or their sponsors wield (McCarthy 1973, 150; Held 1980, 396; see also Thompson 1982, 129). Clearly, in a complete account, the Ideal Speech Situation must be sensitive to these sources of deviation from symmetry. However, Habermas believes that this sensitivity is built into the theory, and he sees the detection of deviations as unproblematic: "Should one party make use of privileged access to weapons, wealth or standing, in order to *wring* agreement from another party through the prospect of sanctions or rewards, no one involved will be in doubt that the presuppositions of argumentation are no longer satisfied" (McCarthy 1973, 150–51; Habermas 1982, 272–73).

The theoretical status of the Ideal Speech Situation needs to be clarified. It is not ideal, that is, it does not enunciate a set of utopian conditions designed to gauge the rationality of any actual interchange. Neither is it a generalization from actual communicative practices, nor a set of rules designed to control events, like those that manage traffic flow. Rather, it is a reconstruction of the criteria that make a certain kind of experience possible, namely, in the case of science, the rationally motivated agreement that must be the foundation of knowledge. Whether its criteria are realized in a particular communicative exchange is irrelevant to the robustness of its theoretical status: together, these criteria form "the constitutive condition of rational speech" (McCarthy 1973, 258–59). This theoretical robustness, however, does not preclude testing the claims of the ideal speech situation against the world. Indeed, such claims must be tested because "rational reconstructions of universal or species competences cannot make the strong aprioristic claims of the Kantian project. They are advanced in a hypothetical attitude and must be checked and revised in the light of the data, which are gathered a posteriori from the actual performances and considered appraisals of competent subjects" (McCarthy 1982, 60). Habermas agrees: "Linguistic reconstruction calls for empirical inquiries undertaken with actual speakers" (qtd. in McCarthy 1982, 61).

Given the persistent misapprehensions of the Ideal Speech Situation, however, misapprehensions that are attributable in part to its misleading name, I would like to change that name for the purposes of this chapter. I will refer not to the Ideal Speech Situation but to the rationally enabled speech orientation, the acronym for which is REASON. This rebaptism for the nonce captures, I believe, the centrality of speaker attitude in any exchange involving communicative action and avoids any implication of philosophical idealism.

This chapter is an attempt to test REASON against an actual case, to define peer review as a rationally motivated system whose goal is the mutual understanding that a particular contribution belongs, at least tentatively, at the advancing edge of a particular science. It is also an attempt to generalize from a single case, a procedure ordinarily impermissible. In this instance, however, a license to generalize is derived from the premise of institutionalization, a status that, I believe, makes plausible the typicality of those aspects of peer review examined here. This exploration of an empirical embodiment of REASON focuses on the contents of one issue of one biological journal. I follow the trail of those papers that initially generated substantial disagreements among the correspondents but were eventually accepted for publication. These disagreements created a rich collection of texts: full cycles of referee criticism, editorial mediation, and authorial rebuttal and repair.

Deviations from REASON

REASON permits each interlocutor an equal opportunity to initiate speech acts. In an obvious way, a review cycle deviates from this principle: after their initial request, authors never initiate speech acts but only respond to those of the referees or the editors. In addition, there is usually only one round of referee critique, editorial mediation, and authorial response, hardly enough to ensure the mutual understanding that must precede rational consensus. Moreover, because interaction between referee and author is generally out of the question, authors are forced to respond without benefit of interactive clarification. They must reply to every question and respond to every criticism, despite the fact that, were interaction possible, some questions might not have been asked, nor some criticisms made.

Habermas's second criterion for REASON is that all interlocutors have the opportunity to reveal their "inner natures," to allow their discourse to become transparent to their full subjectivity. Because referee reports and author responses are frequently full of the emotion-laden language routinely absent from scientific papers, we may infer that peer-review exchanges fulfill this criterion. This inference would be mistaken. Peer-review communication operates under two imperatives: discourse must free of emotion, and it must be polite. If these imperatives are set aside, it is invariably referees who are the first offenders; only then can authors follow suit with impunity. For example, a referee carps:

> I have serious doubts about the numbers unless some moderately magnified, representative micrographs, of good quality, of the different tissues, are furnished. The reader is entitled, at least with the present state of the art and acceptance of stereology [the science of three dimensions] as a quantitative probe, to determine visually, if only intuitively, whether the numerical data may be reliable.

An author replies in kind:

> His reference to the ". . . present state of the art . . . " of electron microscropy is intriguing. Glutaraldehyde-OsO4 TEM [transmission electron microscopy] has been around since before I was in junior high school, and I have never considered it an especially difficult "art." Maybe I have just been blessed with cooperative tissues that prefer to die gracefully??? (The c_____ tissues, I mean, my own personal tissues will go down fighting!)

Whatever its presence in author and referee discourse, emotion is entirely absent from author-editor correspondence in the cases of these papers, papers accepted contingent upon revision. First of all, correspondence of this sort is direct contact with strangers: politeness is the general rule. But if authorial politeness is clearly deferential, editorial politeness is more complex in its motivation. It is doubly involuted: accompanying referee reports, it is a demand disguised as a request and a demand from the editor disguised as a demand from the referees. Note the difference in tone between an editorial letter to the author and its accompanying folder note, meant only for private consumption:

> We are enclosing the comments of two reviewers and one copy of your manuscript. Both reviewers offer many suggestions, which should be carefully considered and incorporated into the revision.

> While both reviewers recommend publication, both reviewers have made so many important and detailed comments and suggestions that only a revised manuscript can be accepted for editing. [signature of one of the editors]

The polite fiction in the letter anticipates and mitigates any authorial wrath and deflects any remaining negative feelings away from the visible editors and toward the anonymous referees.

There are two additional criteria of REASON: all interlocutors must be free to use any speech act, and they must have equal power over the exchange. But, clearly, authors are prohibited from issuing commands and inhibited from asking critical questions, while editors and referees do both freely. This difference is an index of the power of the latter and the relative powerlessness of the former. To submit a paper to a scientific journal is to make a request, a speech act that creates the possibility of denial (Searle 1969, 66). Power is also divided unequally in less obvious ways. According to one study, in physics, one-third of referee judgments were contributed by only twelve per cent of referees, those of higher rank; additionally, their judgments were somewhat harsher than those of their lesser-ranked colleagues (Zuckerman and Merton 1973, 483; 490). These facts are consistent with the notion that referees and authors need not be equally constrained by the prospects of role reversal, the possibility that the judges themselves will be judged by those who once came before them.

Further, the general rule of referee anonymity forces authors to frame a reply in ignorance of their interlocutors. Of course, to balance matters, authors may also be anonymous, but genuine author anonymity in the sciences may well be an ineffectual charade: any worthwhile research program is an intellectual fingerprint. Indeed, the most thorough study of scientific peer review, that of the

National Science Foundation, concludes "that 'blinding' of proposals is both impractical and ill-advised" (Cole and Cole 1981, 56; see also Cole, Rubin, and Cole 1978). Finally, because multiple submission is seen as unethical, authors are forced to endure delays beyond their control, delays that may cause them but not their editors or referees considerable anxiety and that may reduce the value of their submission as a contribution to ongoing scientific debate.

In addition to these criteria, REASON must be open to a progressive radicalization of the arguments that are brought forward in criticism and rebuttal; this means that interlocutors must be free to move up the communications ladder from a given level of discourse to increasingly reflexive levels (McCarthy 1982, 305). As general rule, however, peer review blocks discussions of its own fairness and of the ultimate value of the knowledge to which it gives preliminary certification. For example, editorial decisions are without formal appeal. In fact, any appeal is unusual and must be motivated by exceptional circumstances, such as the allegation of systematic referee bias. In appeal, of course, it is not the decision that is at issue; it is the process by which the decision was reached (Habermas 1979, 64; Pinch 1985, 182).

Finally, the lack of reliability among referee and editor judgments counts against the alleged rationality of the consensus. Though we may concur that referee global judgments are reliable in the sense that two scientists acting independently will usually agree about the approximate value of a paper they are refereeing (Polanyi 1964, 51), nevertheless, we may doubt whether referees would agree concerning the basis of their agreement: in my admittedly small sample, there was consistently little overlap between the questions and comments of two referees of a single paper.

The Correction of Deviation

The criteria for REASON have equality of power as their common denominator. Although peer review in science clearly deviates from this principle, it also attempts to correct for this deviation. In the pursuit of a balance of power between authors and referees, editors generally assume that the rejection of a paper depends on a clear negative decision on the part of all referees, a split decision ordinarily favoring the authors. In conveying the results of peer review, moreover, editors take care to shield authors from the worst of referee excesses. For example, one referee begins his report by labelling the paper's abstract "incomprehensible" and ends with the following peroration:

> The entire paper appears to be a fantasy. The methods used . . . are not at all conducive to understanding the developmental anatomy of this s____. In my opinion, the paper is unscientific and is not worth

> publishing. The paper makes no contribution to the solution of the problem raised in the last paragraph of page 2 or any other problem.

Because they were permitted to savor this report uncensored, the authors could fully appreciate the editors' inventive tact: "Both reviewers offer constructive criticism, and [the other reviewer's] suggestions appear especially helpful."

In the case of another paper, the criticisms of the referee seem devastating:

> The authors contradict themselves on a very important point. . . . [A] more rigorous and thorough analysis of the methodology for enzyme cytochemistry [chemistry of cells] is in order. Since this has not been performed prior to submission of the present manuscript, it is essential that it be done before the manuscript should be accepted for publication. As the manuscript stands, it is not much more than a compilation of a few observations that have limited and questionable application to biological problems. At best, the data represent interesting observations with little significance and limited importance.

The authors, no doubt, would have been upset had they experienced this merciless assault; however, in this case, they received only one, essentially favorable, referee report.

Whatever the motives for these hostile reports—and in one case at least, on the basis of evidence not presented here, conflict of interest may be adduced—they plainly reflect, not on the papers as responsible science but on the referees as responsible members of a scientific community. Like referees who must be badgered to return reports in a timely manner, those who misuse their temporarily bestowed power are likely to be stricken from editorial rolls.

In the case of papers accepted contingent on revision, my chosen subset, referees, originally retained as judges, are redefined by editors as repairers and collaborators; authors, though still defendants, have a new, editorially defined task: the reestablishment of their authority against referee requests for alteration and amplification.

This change in relationship does not mean that authors must comply fully with referee requests. Certainly, authors must consider these carefully, but the admitted failure to provide satisfactory answers to referee questions is no bar to publication. For example, a referee asks: "Is cell death involved in the formation of some portions of the hollow s____?" In their reply, the authors confess their failure to find a satisfactory answer:

> We have gone through our slides once again to find out if death is involved in the formation of some portion of the hollow s_____.

> . . . However, we could not get enough information to make a categorical comment on this aspect. We have, therefore, not included [the] cell degeneration aspect in the manuscript.

Indeed, publication can proceed even when there is a serious disagreement with a referee concerning the fundamental assumptions of the field. For instance, a referee provides support for the contention that "the systematics section [of this paper] . . . needs some rethinking":

> For historical reasons and due to the thorough studies on Ar________ and Rei_______ by the author he would like to erect the order Ir__________ in which it is proposed to include Ib____ . Ib_____, however, [is] known as a whole ______ including external h____, sp________, etc.

The author responds by roundly attacking this last referee assumption:

> Although I do admit that Ib____ is better known than the other genera in terms of overall morphology, it is very far from being completely understood as a whole _____. S_____ and B_____ (1973, 368) were careful to point out the limitations of their evidence on external morphology especially in relation to the poorly preserved structures thought to be sp______. Essentially nothing is known about the histology of these structures, and on the basis of the evidence presented in their paper, even the occurrence of sp_______ can hardly be accepted as irrefutable fact. Information on the anatomy of Ib____ is even more fragmentary. In fact, I wouldn't at all be surprised that when more is known internally about Ib_____, it will be shown to be synonymous with Ar__________ which at present is far better known internally.

In author responses, we see two central strategies of defense. In the excerpts just quoted, authors disagree by plausible counterargument, citing, as do lawyers, whatever authorities suit their purpose, including the authority of re-observation. In a second strategy, authors side, wherever possible, with the authority of the editors or of the positive referee. In this strategy, the absence of referee consensus is incorporated in rational consensus formation; it becomes a source for defensive argumentation. Concerning a referee request for additional electron micrographs, an author says:

> Nothing infuriates me more than published unnecessary TEMs [transmission electron micrographs] . . . . If you, as editors, really

> want them, I will supply them in quantity, but otherwise I think they are a waste of space.

Another author summarizes as follows:

> I can only conclude that Reviewer A objects either to my inclusion of taxa on the basis of unique and derived features, or to my use of a new name, or both. To be perfectly honest, I am greatly surprised by the difference in the response of reviewers A and B on precisely these points. My approach is understandable in the paper, I believe, and reasonable.

But dissent is far from the rule, because dissent risks rejection. Ordinarily, authors reestablish their authority by treating referees as collaborators, as additional, if invisible, authors of the paper; indeed, sometimes, they acknowledge in print the help of these still-anonymous critics. Authors also comply patiently with requests as time-consuming as an omnibus reexamination of slides; they back down gracefully in the face of error and willingly curb unwarranted speculations to the extent, in one case, of creating a new major division, as one author did: "A Limitations and Recommendations section was added to explain the limitations of our study and summarize the findings."

Editors are themselves limited in power by the professional situation in which they find themselves. All wish to publish journals that will be widely read and frequently cited. Though they do not rely on any one author, they do depend on authors generally to supply a steady stream of acceptable manuscripts. Equally, authors do not rely on one editor's decision; they have the power of the free market for ideas in their field of study. So long as there exist other journals of approximately equal prestige to which rejected authors may send their papers, editors are faced with the prospect of seeing in print or, worse, in the Science Citation Index, papers they judged unsalvageable for their journal. None of these considerations is decisive in any particular case; however, in concert, such considerations set serious limits on the irresponsible use of editorial power.

Necessary Deviations

In Habermasian terms, peer review in science is a system of communication that achieves rational consensus by means of the balance it strikes between deviations from REASON and their correction. Because these deviations prevent the full realization of REASON, it might seem that fully rational consensus depends on their total elimination: in an ideal society, scientific peer review would permit unimpeded author initiative, endless rounds of give and take, and unchecked openness among authors, editors, and referees. Until these circumstances obtain, peer-review communication can not fully realize rational consensus. But peer review is an institutionalization of REASON, a practical matter: the question is not whether it embodies a full realization of REASON but whether it "can be justified as leading to outcomes that would, ideally, be agreed to in unconstrained dialogue" (McCarthy 1987).

This is not to deny the initial attractiveness of the full realization of REASON in peer review, a realization that would ensure both the quality of accepted papers and the general intellectual and political health of the community. At the same time, this realization would guarantee that all potentially worthwhile research results would be shared with appropriate peers, that proper credit would be given, and that a scarce resource—journal space—would be responsibly allocated. Unfortunately, placed back in its proper context, the form of life of science of which it is a part, the pursuit of fully rational consensus competes with an equally important good, an equally scarce resource—professional time.

The need to conserve professional time is an imperative in every discipline; it explains the strong constraint against multiple journal submissions of the same paper; it also accounts for the powerful resistance to otherwise sensible suggestions for peer-review improvement, for example, the suggestion that referee reports be shared with authors *before* the editorial rejection of a paper. A decisive movement in the direction of REASON would improve an author's position only at the expense of the professional time of editors and referees. By substituting review for research and self-reflection for science, such a movement would also impede the creation and the timely communication of potentially valuable science.

The full realization of REASON is necessarily limited. This limit stems from a compromise, always more or less uneasy, among principles that cannot coexist in their more extreme forms. REASON cannot be viewed as a goal toward which we gradually move, because such a construal involves us in paradox. We must suppose that rational consensus is a good we can pursue without regard to its possibly undesirable effects on other equally desirable goods, such as the production of new science. In other words, we must subscribe to the view that all good things are compatible, that in an ideal world, all good things will form an integrated whole (Berlin 1978, 102, 95): we must subscribe to an Enlightenment myth.

For Habermas, however, particular social institutions such as peer review are practical matters, necessarily open to compromise:

> It is a question of finding arrangements which can ground the presumption that the basic institutions of the society and the basic political decisions would meet with the unforced agreement of all those involved, if they could participate, as free and equal, in discursive will- formation. (1979, 186)

The heedless pursuit of rational consensus would have an unavoidable and undesirable cost: REASON cannot be fully realized without devaluing professional time and progressively undermining peer review's central purpose: the timely initial certification of new science. For this reason, every time a new scientific or scholarly journal is founded, peer review is reinvented with virtually identical constraints and with the virtually the same balance between deviations from REASON and their correction.

Peer Review, Publication, and Scientific Knowledge

Scientific reports and peer-review documents are very different communicative enterprises. Scientific reports are primarily cognitive; they are about truth, the link between statements and the world; their dominant speech act is the confirmative, an expression by which writers indicate that their beliefs come about as a result of some truth-seeking practice (Bach and Harnish 1979, 46). In contrast, peer-review documents are primarily about trust, a network linking authors, editors, and referees in a community of faith. In peer review, then, the results of a cognitive process are certified, at least initially, by one based on trust. This certification necessarily diminishes the status of the scientific knowledge it certifies: cognitive claims are judged not by testing against the world but procedurally by consensus. The knowledge peer review certifies is grounded in intersubjective argument.

After peer review, claims that have been vigorously disputed in camera are turned again toward the world as meant by science; after peer review, the machinery of peer review disappears from sight. Publication performs this task; it frees the scientific report from the doubts that peer review generates, those that shift its status from knowledge about the world to knowledge about the agreement among referees. Publication obliterates all traces of the procedure by which the knowledge it asserts is certified; as a consequence of this obliteration, publication renews the credibility peer review undermines. The candidate knowledge in published papers is now ready to be tested against broader consensuses of scientists and, more importantly, against the world.

Once they have served their purpose, peer-review documents become generally unavailable to scrutiny. Scientific reports are read only in the context of other reports—all of which partake in a struggle for conceptual existence, a survival of the fittest theories, those that exhibit the closest correspondence between words and the world. This is as it should be: the superiority of science as a way of knowing rests not only on the consensus produced by REASON but also on truth-seeking practices, one of which is the capacity of experimental programs to confirm or disconfirm theories. It is these practices that constitute the objective side of scientific certainty.

Chapter 9 explores the closure that such experimental programs bring. In the case of cold fusion, we will see the disconfirmation of a theory that, if true, would have undermined a powerful existing consensus; in the case of the experimental violation of Bell's inequality at the quantum level, we will see science disconfirming not only a theory but also a philosophical position. Before taking up these issues, however, I need to explore how candidate knowledge spreads through a scientific community at a time when its claims are still seriously underdetermined by the available evidence in their favor. My example will be the spread of Copernicanism, the subject of the next chapter.

8 The Spread of Scientific Knowledge

The Initial Acceptance of Heliocentricity

Copernicanism was slow to spread, inhibited in part by the defects in the Copernican system itself; at least at its beginnings, the Copernican system was no better than the Ptolemaic as a tool for practical astronomy. Still, these early deficiencies, well-known to modern historians, hardly justify Paul Feyerabend's view in *Against Method* that allegiance to the Copernican system in Galileo's day amounted to "a blind faith," supported not by evidence and arguments but "*by irrational means* such as propaganda, emotion, *ad hoc* hypotheses, and appeal to prejudices of all kinds" and by "Galileo's trickery." Yet, Feyerabend's views on the Copernican revolution cannot be lightly dismissed. He has a firm grasp of its essential condition: "an entirely new world view containing a new view of man and of his capacities for knowing." He is also acutely aware "that developing a good theory is a complex process that has to start modestly and that takes time" (1978, 153–54, emphasis in original; 84; 152, emphasis omitted; 98).

Without endorsing Feyerabend's radical notion that the development of science is and ought to be essentially anarchic and fundamentally irrational—without endorsing Feyerabend's own rhetoric—it is possible fully to endorse—and, indeed, usefully to extend—one of his deepest and truest insights: that early in the development of a new scientific world view, at a time when a new theory is seriously underdetermined by the evidence, "style, elegance of expression, simplicity of presentation, tension of plot and narrative, and seductiveness of content become important features of our knowledge" (1978, 157) Applied not to Galileo's *Dialogue* of 1632 but to the first published work on Copernicanism, Rheticus's *Narratio Prima*, published in 1540, three years before *De Revolutionibus*, Feyerabend's insight can show how and why rhetoric operates at the inception of at least one scientific revolution and how it initiates the social network of scientists who believe, which final acceptance requires.

In this chapter, I shall show how in *Narratio Prima*, Rheticus justifies his choice of the Copernican over the Ptolemaic hypothesis. To do so, he goes beyond argument and evidence. Naturally, he gives reasons his training leads him to believe are scientific. But because as a result of this very training, he cannot regard these reasons as sufficient for full conviction, he sets them within the frame of a conversion narrative: he unites argument and narrative into a single structure, a model for radical intellectual change.

This model, which I call *rational conversion*, was an ingredient essential to the spread of Copernicanism in the late sixteenth and early seventeenth centuries. The new Copernicans shared with their master an austere and intransigent realism: the new astronomy must be a mathematically parsimonious system in exact agreement with accurate observations and in precise conformity with a correct physics. But the realization of this explanatory ideal was not possible until the advent of classical dynamics over a century after Copernicus's death. Thus, in a period of transition, the model of rational conversion made the shift to Copernicanism psychologically tolerable: it facilitated change.

At the same time, as the work of Tycho Brahe, Michael Maestlin, and Johannes Kepler makes clear, the model of rational conversion facilitated an apparently contrary purpose: the undermining of its own justificatory role. Because the narrative element in rational conversion was by definition unscientific, it became an irritant, a psychic itch that provided the continued impetus for the eventual abandonment of all reliance on psychological models undergirding the espousal of the new paradigm. It was not until Isaac Newton that the task was completed to the satisfaction of a sizable segment of the then scientific world; only then did astronomy rest, in these scientists' view, on a fully scientific base. On this reading, *Narratio Prima* becomes the opening move in an eventually successful campaign to create the first modern exact science.

The Scientific Case for Heliocentricity

The problem Copernicus faced was the same as that of his great predecessor, Ptolemy: to make coherent physical sense of the apparent behavior of celestial objects. The sun rises and sets, and, on occasion, there is a solar eclipse; the moon has phases, and, on occasion, there is a lunar eclipse; some of the planets exhibit retrograde motion (that is, at times they seem to move backward); the stars apparently do not exhibit parallax (that is, they do not seem to change their apparent position, allowing the inference that the Earth is not changing its position relative to them); the comets dart through the night sky for a while and seem to disappear again, sometimes to return after a number of years. We now know the reasons for all these appearances lie in the actual configuration of our solar system and of the universe of which it is a part: for example, the planets, including the Earth, move in elliptical orbits around the sun, though these orbits are not on the same plane; the stars are impossibly far away—farther than even Copernicus dreamed; the Earth rotates on its axis at about one thousand miles an hour and, what is more, wobbles a bit; the whole universe is expanding, its various objects moving very rapidly apart from one another. Before Copernicus, none of this was part of astronomical knowledge; after him, however, it was virtually impossible to avoid engagement with his two major conclusions: the Earth rotates on its axis, and it revolves in an orbit around the sun.

The radical nature of Copernicus's two major conclusions should not blind us to the conservative origin of his methods. His science is a field of study with a long history, many of whose components are shared with his Ptolemaic fellows and predecessors. To describe these components is largely a matter of identifying those statements that all astronomers took as self-evident. Copernicus shares with his scientific past three sorts of self-evidence: an astronomical theory must reason by means of Euclidian geometry, it must agree with the most accurate astronomical observations, and it must conform to a correct—essentially an Aristotelian—physics.

In *Narratio Prima*, therefore, Euclidian methods are central: "my teacher . . . has embraced the whole of astronomy, stating and proving individual propositions mathematically and by the geometrical method" (1959, 109–10). But geometric reasoning is not sufficient: there must be exact agreement between theoretical conclusions and accurate observations. Concerning a discovery of his teacher, Rheticus says that it "is indeed worthy of the highest admiration, since it is achieved with such great and remarkable agreement" with observations (1959, 121). By means of geometry, then, Copernicus intends to harmonize in one system all of the observations available to him—both those he has made himself, and those he has inherited from such great predecessors as Ptolemy (1959, 131–32).

But geometry and accurate observation are in themselves insufficient. Rheticus cites with approval Aristotle's distinction between the mathematician and the physicist, between a person concerned with abstract dimensions and quantities and one concerned with these only insofar as they are the properties of real bodies (1959, 140; Aristotle 1929, 1:116–26.). Because the astronomer deals with the properties of real bodies, "the mathematician who studies the motions of the stars is surely like a blind man who, with only a staff to guide him, must make a great, endless, hazardous journey that winds through innumerable desolate places. . . . God will permit him to try his strength for a period of years, that he may in the end learn that he cannot be rescued from threatening danger by his staff" (1959, 163–64). Thus, it is only through its union with a correct physics that mathematics coupled with observation realizes its full power as a tool of physical astronomy, its ability to reflect in its own simplicity the actual simplicity of the universe that it describes: "Lest any of the motions attributed to the earth should seem to be supported by insufficient evidence, our wise Maker expressly provided that they should all be observed equally perceptibly in the apparent motions of all the planets; with so few motions was it feasible to satisfy most of the necessary phenomena of nature" (1959, 161; see also 137, 149).

It is Rheticus's point that the assumption of heliocentricity permits the parsimonious demonstration of astronomical truth: "If anyone desires to look either to the principal end of astronomy and the order and harmony of the system of spheres or to ease and elegance and a complete explanation of the causes of the phenomena, by the assumption of no other hypothesis [than heliocentrism] will he demonstrate the apparent motions of the remaining planets more neatly and correctly" (1959, 164–65). In fact, "the assumption of the motion of the earth on an eccentric [leads to] a sure theory of celestial phenomena, in which no change should be made without at the same time re-establishing the entire system, as would be fitting, once more on proper ground" (1959, 140). Because the heliocentric hypothesis yields astronomical truth so readily, it is also the primary candidate for the central physical truth about the universe, a truth so overwhelming that "it seems impossible that the earth should occupy the center" (1959, 137).

Copernican astronomy has now met the goal of science as set out in the *Posterior Analytics*: "The hypotheses of my teacher agree so well with the phenomena that they can be mutually interchanged, like a good definition and the thing defined" (1959, 186; Ross 1971, 49–54; Aristotle 1975, 64–65). With mathematics and physics as his guides, Copernicus has taken Plato's admonition seriously: he has ensured a "clear waking vision" for astronomy by giving an "account" of the "assumptions" of his system (Rheticus 1959, 142). We now have a total system "in agreement with the truth" (1959, 193).

The Rhetorical Case for Heliocentricity

Brahe's *Astronomae Instauratae Progymnastmata*, Johann Schöner's *Opusculum Astrologicum*, Christoph Clavius's *In Sphaeram Joannis de Sacro Bosco Commentarius*, *Astrolabium*, and *Gnomonices*, Girolamo Fracastoro's *Homocentrica, sive De Stellis*, the anonymous summary of Georg von Peurbach's *Novae Theoreticae Planetarum*—the form of these typical Renaissance astronomical treatises is essentially that of Ptolemy's *Almagest*, itself clearly derived from Euclid. Copernicus's *Commentariolus* also has a transparently Euclidian form, and *De Revolutionibus* is plainly imitative of the *Almagest*. But the overarching structure of *Narratio Prima* is story-like: the scientific case for Copernicanism is embedded in a framework of conversion.

Initially, Rheticus presents Copernicus, his protagonist, as a great traditional astronomer, steeped in Ptolemy. Accordingly, in the first third of *Narratio Prima*, the motion of the Earth goes unmentioned, and those aspects of Copernicus's astronomy are dwelt upon that do not require a discussion of heliocentricity, such matters as the motion of the fixed stars, the length of the solar year, and the obliquity of the ecliptic (the apparent annual path of the sun and the planets among the stars). In these pages, Copernicus is presented as the heir to Ptolemy, his work written "in imitation of" the earlier astronomer (1959, 131). As Ptolemy's heir, Copernicus relies on the most accurate observations—those he had made himself and those of his predecessors, especially Ptolemy—as a basis for an improved theory. In the first third of *Narratio Prima*, indeed, Copernicus is Ptolemy's partner in the restoration of their mutual science (1982, 42).

Ptolemy, the great predecessor, is mentioned over fifty times, often with great praise:

> Ptolemy's tireless diligence in calculating, his almost superhuman accuracy in observing, his truly divine procedure in examining and investigating all the motions and appearances, and finally his completely consistent method of statement and proof cannot be sufficiently admired and praised by anyone to whom Urania is gracious. (1959, 131)

But Rheticus's respect for his teacher exceeds his respect for Ptolemy, for his teacher has "a burden greater than Ptolemy's," the burden of incorporating intervening astronomical observations into his system. Repetitively referred to by Latin honorifics for teacher—*D[ominus] Praeceptor, D[ominus] Doctor Praeceptor*, or *D[ominus] Doctor* (Ptolemy is referred to only by his bare name)—Copernicus creates a monument to observational astronomy, astronomical tables of really lasting value (1959, 131; 126).

In the midst of this portrayal of Copernicanism as the natural outgrowth of Ptolemaic astronomy, the crucial Ptolemaic assumption of geocentricity is suddenly and completely discarded. Near the end of a section on lunar theory, Rheticus says: "These phenomena, besides being ascribed to the planets, can be explained as my teacher shows, by a regular motion of the spherical earth; that is, by having the sun occupy the center of the universe, while the earth revolves instead of the sun on the eccentric" (1959, 135). Because it comes without warning in a text up to then orthodox, this sentence was certainly a surprise to its sixteenth-century readers. It was also a shock: to believe it was to turn one's back on the central tenet of years of training, to discard completely the unquestioned assumption of respected colleagues and admired teachers, to desert forever the cosmos of fourteen centuries of astronomy. Little wonder that for over a century, Copernicanism had relatively few significant adherents.

Sudden as this shift is, it is not without narrative preparation and does not long remain without argumentative justification: the conflict between Copernican and

Ptolemaic astronomy is presented as a conflict within the scientific tradition and within Copernicus himself. While "walking in the footsteps of Ptolemy," Rheticus says, Copernicus nevertheless "became aware that the phenomena, which control the astronomer, combined with mathematics to compel him to make certain assumptions even against his wishes." He "had to assume" his hypotheses; he had to reach his conclusions "in obedience to the command given by the observations" (1959, 186; translation emended, 140, 151). But, however radical, these were conclusions with which Aristotle would certainly have agreed: "I am convinced that Aristotle, who wrote careful discussions of the heavy and the light, circular motion, and the motion and rest of the earth, if he could hear the reasons for the new hypotheses, would doubtless honestly acknowledge what he proved in these discussions, and what he assumed as unproved principle" (1959, 142).

In an authentic search for truth, it is these unproved principles which, according to Aristotle's teacher, Plato, must be examined (1959, 142): to deny the license to question geocentricity is to deny a central principle of the intellectual tradition in which Ptolemy worked, a denial that makes impossible any honest search for the physical reality behind celestial appearances.

In telling Copernicus's story, Rheticus is as sensitive to the value of contrast as is his great predecessor in conversion narrative, the author of *Acts* in the New Testament. The drama of rational conversion, which takes Copernicus from essentially the same observations as Ptolemy to a diametrically opposed central conclusion, is intensified by the master's innate conservatism. The espousal of heliocentrism is especially reluctant, because "he is far from thinking that he should rashly depart, in a lust for novelty, from the sound opinions of the ancient philosophers, except for good reasons and when the facts themselves coerce him" (1959, 187). (The parallel with Darwin is irresistible, a conservative thinker led to radical conclusions by the force of "certain facts" with which he was "much struck.") Even after Copernicus himself was entirely convinced of heliocentricity, we learn, he was reluctant to publish his results, lest others be upset. Only the repeated urging of the Bishop of Kulm that there was no place for secrecy in science finally persuaded the great astronomer to publish not merely his tables but his whole system with proofs (1959, 192–93).

In recounting Copernicus's conversion to heliocentrism, Rheticus is also recounting his own: "In my teacher's revival of astronomy I see . . . with both eyes and as though a fog had lifted and the sky were now clear" (1959, 168; Westman 1975b). Throughout *Narratio Prima*, Rheticus's own faith in his teacher's hypothesis increases slowly, a pace mirrored in his slowly increasing assertiveness. At first, a notion by which the phenomena merely "can be explained," heliocentricity is soon afterward entertained more seriously: "Should we not attribute to God, the creator of nature, that skill which we observe in the common makers of clocks?"

But the idea is still treated with a certain tentativeness: the argument is still full of "can," "suppose," and "as it would if" (1959, 135; 137; see also 145, 150). No such hesitancy, however, attaches to this pronouncement near the end of the work and embedded in a discussion of the intersections of Venus with the ecliptic: "That this covenant of earth and planet might be everlasting, God ordained that the first small circle of libration [real or apparent oscillation] . . . should revolve once in the time in which one return of Venus to either of the movable nodes [intersections] occurs" (1959, 184; see also 185).

This increase in Rheticus's conviction is also mirrored in the gradual elevation of Copernicus's symbolic status: from Ptolemy's heir, to king, to general, to philosopher, to mythical hero: like Atlas, shouldering the world or, like Orpheus, rescuing the muse of astronomy from the underworld (1959, 131, 132, 150, 162–63, 164). Such increasing mythical status is entirely appropriate, for rational conversion has led only to the discovery of the Neoplatonic truth: "self-caused," the sun is "the source of all motion and light"; "God stationed in the center of the stage His governor of nature, king of the entire universe, conspicuous by its divine splendor" (1959, 139, 146, 143; Kuhn 1957, 128–33).

By this time, it should be clear to readers that they must reread or remember anew the first portion of *Narratio Prima*. Crucial words—terms such as *hypothesis, theory, cause*, and *truth*—have changed their meaning. For example, a few pages before Rheticus's abrupt revelation of Copernicus's heliocentricism, this passage occurs:

> Nevertheless, the observations of all scholars and heaven itself and mathematical reasoning convince us that Ptolemy's hypotheses and those commonly accepted do not suffice to establish the perpetual and consistent connection and harmony of celestial phenomena and to formulate that harmony in tables and rules. It was therefore necessary for my teacher to devise new hypotheses, by the assumption of which he might geometrically and arithmetically deduce with sound logic systems of motion like those which the ancients and Ptolemy . . . once perceived . . . and which careful observations reveal as existing in the heavens to those today who study the remains of the ancients. (1959, 132)

In its original context, before the heliocentric revelation, this statement about new hypotheses can refer only to the technical innovations presented in the first third of *Narratio Prima*. The heliocentric revelation having occurred, the passage concerning the necessity of new hypotheses must be read rather as an announcement of a different order in the heavens, an order dictated by necessity (1959, 132). Of his teacher, envisioned in the first third of *Narratio Prima* as the ruler of the kingdom of astronomy, Rheticus says: "May he . . . deign to govern, guard, and increase it, to the restoration of astronomic truth" (1959, 131). By the end of *Narratio Prima*, astronomical truth has been redefined, and hypotheses have become what they are throughout *De Revolutionibus*: physical truths about the universe.

The Need for Rational Conversion

To explain his master's espousal of heliocentrism and to justify his own adherence, Rheticus invented rational conversion. He did so because the scientific arguments for Copernicanism were still weak. Until Brahe, Kepler, and Newton, the heliocentric view fell short of its own highest goal: "to explain completely," as Copernicus says in the *Commentariolus*, "the structure and motions of the universe" (1959, 90; Prowe 1967, 2:202; my translation).

To Copernicus, and therefore to Rheticus, to explain completely had at least two meanings. First, it meant a thoroughgoing realism: astronomers must describe the unique, that is, the actual lunar and planetary orbits. On this reading, the Copernican revolution consisted not in heliocentricity only but also, and perhaps primarily, in this insistent realism: "In this sense—if in no other—Copernicus may be considered the first great figure of the Scientific Revolution. It was essentially his attitude which came to prevail with Kepler, Galileo, Descartes, and Newton" (Grant 1962, 215–16). Although *De Revolutionibus* was preceded by a preface (later shown to be written by Osiander) disavowing this view, no one studying the work with any care could have been deceived; certainly, the Congregation of the Index was not. In 1620, it approved the work only when appropriately emended: when Copernicus spoke of *demonstrating* the triple motion of the Earth, the Congregation insisted that he speak only of the *hypothesis* of that motion (De Morgan 1954, 95).

But Copernicus could not fulfill his realist program. His epicycles and eccentrics—intricate combinations of uniform circular motions—could never describe unique orbits. Not because epicycles and eccentrics could not be real; indeed, Averroes is criticized by Rheticus for his conclusion "that epicycles and eccentrics could not possibly exist in the realm of nature" (1959, 194). Instead, the problem was that epicycles and eccentrics combined to create too many solutions to lunar and planetary orbits. All were mathematically equivalent; consequently, all had an equal claim on belief.

Copernicus was well aware of this equivalence (Rheticus 1959, 74n.; see also 138, 168). Only one of these solutions could be correct, yet it was impossible to choose the actual among them—if any among them was actual. But if a real choice was impossible, the Copernican model for the universe was, in this respect at least, a mere calculating device: "What has been determined cannot have innumerable explanations; just as, if a circumference is drawn through three given points not on a straight line, we cannot draw another circumference greater or smaller than the one first drawn" (Copernicus 1959, 100–101). Although heliocentricity was real for Copernicus, and epicycles and eccentrics could be real, *his* epicycles and eccentrics remained mathematical conveniences. A passage from *De Revolutionibus* makes clear Copernicus's mature view in this matter, a blend of skepticism and hope: "Since so many arrangements lead to the same result, I would not readily say which one is real, except that the perpetual agreement of the computations and the phenomena compels the belief that it is one of them" (1978, 164).

To Copernicus, and therefore to Rheticus, to explain completely was also to explain simply. In *Narratio Prima*, early in his defense of heliocentricity, Rheticus entertains parsimony as an explanatory value: "This one motion of the earth satisfies an almost infinite number of appearances" (1959, 137). Later in his argument, Rheticus goes further: he assumes parsimony as an explanatory value: "With so few motions was it feasible to satisfy most of the necessary phenomena of nature" (1959, 161; see also 137, 149). But Copernicus's world system remains as he left it, a scheme far from simple, one no simpler than its Ptolemaic rival. True, as Copernicus illustrates it in *De Revolutionibus*, his universe seems simple enough, a system of seven nested spheres (an eighth forms the moon's sphere), but this is not an accurate picture: there are far too few circles. The drawing derives not from Copernicus's actual system but from a class of world-system depictions of which Peter Apian's much-reprinted geocentric universe is a typical example (see fig. 8.1) (Armitage 1962, 80). In Copernicus's illustration (see fig. 8.2), the earlier drawing is redrawn with the positions of the sun and the Earth simply reversed. Copernicus's diagram represents an act of faith in the physical possibility of a heliocentric system.

If to explain completely meant to provide a unique solution to lunar and planetary orbits, one that described them simply and accurately as they actually were, Copernicus fell short of his announced purpose in historical retrospect as well as in his own terms. For Rheticus and for the early Copernicans, therefore, the change to heliocentricity was necessarily a movement of the will as well as the reason. Rheticus's invention of rational conversion to justify change acknowledged the obvious truth, and, in *Narratio Prima*, form followed function.

A Copernican Social Network Forms

When we see in *Narratio Prima* just how far short of his goals Copernicus fell, we may be tempted to argue against the very notion of a Copernican revolution, a position shared by I. Bernard Cohen and Arthur Koestler. In Cohen's words: "The notion that a revolution in astronomy attended the publication of Copernicus's

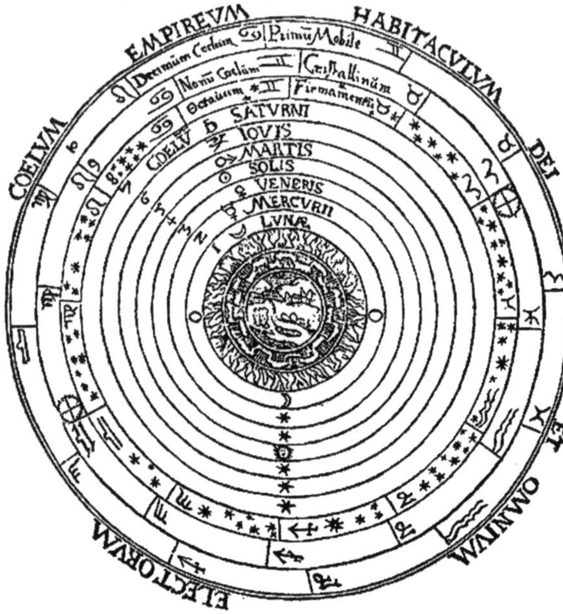

Fig. 8.1. The Ptolemaic Universe

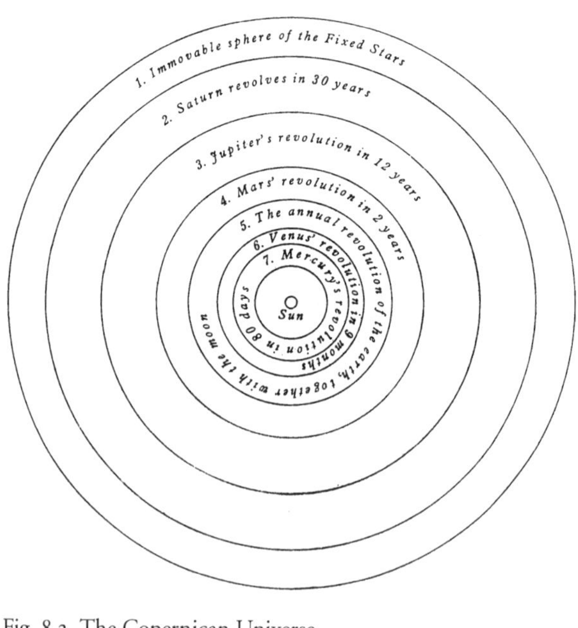

Fig. 8.2. The Copernican Universe

De Revolutionibus in 1543 was a fanciful invention of eighteenth-century historians of astronomy" (Cohen 1985, x; see also 40–125).

The evidence is certainly solid concerning the lack of immediate impact for Copernicanism: sixty-six years separate Copernicus's *De Revolutionibus* from Kepler's *Astronomia Nova*, the work in which, by announcing his first two planetary laws, he inaugurates a new age of mathematical astronomy (Berry 1961, 179–97). During those intervening years, the influence of Copernicus on most professional astronomers was clearly negligible. Indeed, Copernicanism was not accepted in England until 1650 or in Sweden, Spain, Hungary, or Poland until nearly the middle of the eighteenth century (*Colloquia Copernicana* 1, 1972). Although this opposition could be and often was irrational, it could be and often was understandable and defensible, because the arguments for heliocentricity remained less than compelling. On this view, by the time Copernicanism was generally accepted, it is better described as a result of an evolutionary change.

However, this view overlooks important evidence. Koestler calls *De Revolutionibus* "the book that nobody read" (1968, 191; emphasis omitted); to Cohen, Copernicus's masterpiece "had no fundamental impact on astronomy until after 1609" (1985, 38). This scarcely seems fair to a work that Rheticus devoured, Brahe knew well, Kepler studied assiduously, and Maestlin annotated over a period of fifty years (Moesgaard 1972; Kepler 1981, 17; Westman 1975a). These men formed a social network of influence. Maestlin was Kepler's teacher, and Brahe his mentor; it was, in fact, Brahe who held the empirical key to the laws of planetary motion. When Kepler's *Mysterium Cosmographicum* appeared in 1596, Maestlin, who was closely involved in its publication, on his own authority added Rheticus's *Narratio Prima* as an appendix. These connections belie the view that there was no early Copernican impact on astronomy; instead, they support the contention of Robert S. Westman that in the 1570s "a small, but significant group of mathematical astronomers had familiarized themselves profoundly with the technical aspects of Copernican astronomy and granted a serious hearing to its realist claims" (1975a, 54).

To trace this social network of influence is to chronicle a progress that travels from the birth of modern heliocentricity to the threshold of its first significant intellectual flowering: Kepler's three laws of planetary motion. There are verbal links that form a chain of descent from Rheticus to Maestlin (Kepler's teacher) and from Rheticus and Maestlin to Kepler himself, a chain, of rational conversions. Maestlin, who refers to Copernicus as "the Prince of Astronomers after Ptolemy," supports his commitment to the Copernican hypothesis as follows: "I do not wish to approve it on the grounds that I have been deceived and fascinated by love of novelty, but rather, compelled by necessity, I came to it with reluctance" (Westman 1972, 9, 17). This statement has a familiar ring. Indeed, it paraphrases a passage from *Narratio Prima* (1959, 187; see 116 of the present volume). But the influence of Rheticus on Maestlin is also readily apparent from his testimony in the *Mysterium Cosmographicum* (1981, 83–85).

Kepler is another link in the chain. The prefatory material to his *Mysterium Cosmographicum*, his eccentric first masterpiece, clearly documents Maestlin's influence as a Copernican teacher—"the distinguished Michael Maestlin at Tübingen." More to the point, it plainly attests to Kepler's rational conversion. First, he was "disturbed by the many difficulties of the usual conception of the universe" and "collected together . . . the advantages which Copernicus has mathematically over Ptolemy." Next, to prepare for a new teaching position, he was forced into "further contemplation." Finally, he experienced "a strong desire . . . [and] threw [himself] with the whole force of [his] mind into this subject" (1981, 62–63). Kepler explicitly calls this autobiographical process a complete conversion which, nevertheless, was not undergone rashly (1981, 78–79).

Rational conversion to heliocentric astronomy was more than a psychological bridge between intellectual systems. The conflict between Copernicus's uncompromising realism and the deficiencies of the actual system he developed had a distinctly positive force: it acted as a century-long heuristic, inspiring a line of scientists that stretched from Brahe to Newton. Drawing their inspiration from this source, these scientists worked to create a universe they judged structurally and dynamically adequate: a mathematically simple and harmonious total system in accord with the laws of a new physics and precisely descriptive and predictive of celestial appearances, a system that, in their view, accurately described the actual motions of real celestial objects. It was out of the efforts of Copernicus's successors, then, that there emerged a heliocentric universe far different from that of Copernicus but one more nearly in conformity with the explanatory ideal first asserted publicly in *Narratio Prima*.

The place of rational conversion in the initiation of scientific revolutions is not limited to the Copernican. In 1944, Oswald T. Avery and his collaborators published a paper containing a revolutionary conclusion: that genes were made of DNA. Concerning this discovery, Nobel laureate Salvador F. Luria reflects:

> No pressure or competition or sudden emergence of irrefutable data precipitated the decision to publish. In retrospect it appears that at this point Avery may have decided to exert the prerogative of the experienced leader. He exhibited the assertiveness that comes from the habit of success: a willingness to impose on a still confused mass of data a certainty that is emotional as well as rational. Such a source of certainty in science is unrecognized by those who believe certainty comes only after innumerable controls and attempts to disprove. The

> certainty Avery exhibited is more akin to illumination, a sudden vision projecting the possibility of an intellectual leap. (1986, 29)

Later, Luria makes explicit the connection between the revolutions in biology and in physics: Avery published his results as "*a fact*—even though a fact validated as much by conviction as by hard evidence, a fact less fully convincing to experimentally minded critics than Kepler's laws might have been to Huygens and Newton" (1986, 30; emphasis in original). That Avery's faith was justified in a few short years by James D. Watson and Francis H. C. Crick's discovery of the double helix must not obscure the parallel between Avery's case and Copernicus's. Rational conversion is a general model for scientific change in which faith in novelty must precede its firm evidential base if such a base is ever to be revealed.

9 The Incorporation of Claims into Practice

Closure in Science and Its Philosophy

What is the proper relationship between science and its philosophy, the self-reflection that accompanied the birth of science and continues unabated to this day? Philosopher Richard Rorty uses pragmatism to attack what he takes to be two misconstruals of this relationship:

> The scientistic side of pragmatism . . . debunks [the] image of philosophy as a superscience, employing a privileged method and thus attaining a kind of knowledge not available to the mere natural scientist. The holistic side of pragmatism . . . wants to avoid having the natural scientist step into the cultural role which the philosopher-as-superscientist vacated. . . . It wants that cultural role to *remain* unfilled. (1991a, 75; emphasis in original)

From Rorty's perspective, then, no epistemological superiority can be claimed for either discipline, philosophy or science. Moreover, the notion of epistemological ranking is itself spurious and ought to be discarded. On closer examination, however, the relationships between science and its philosophy are more

---

Research note: The conclusions of this chapter are based on a study of the Cold Fusion Archives at Cornell University and the University of Utah and on interviews conducted in Salt Lake City, Utah, with participants from Brigham Young University and the University of Utah. Created by Cornell historian Bruce Lewenstein, the Cornell Archive consists of papers, preprints (not yet published papers), e-mail on computer disks, videotapes of television news programs, peer-review protocols, cartoons, T-shirts, and, most important of all, interviews with key scientists, though not, unfortunately, with Stanley Pons or Martin Fleischmann. The Utah Archive consists of correspondence and memoranda, though it has in all likelihood been purged. Where permission had been given, real names have been used; otherwise, pseudonyms have been substituted. All quotations, except where noted otherwise, are from the Cornell Archive.

resistant to change than Rorty envisions. The superiority of science as a way of knowing may be more than a matter of disciplinary politics, more than a misconception that pragmatism can correct. To explore this claim of epistemic superiority, I focus in this chapter on a point crucial to all intellectual dispute: closure, that is, the decision to halt debate and the judgment that incorporates mere candidate knowledge into knowledge accepted by virtually entire communities of practitioners. I look at two cases: cold fusion, in which science solves one of its own problems, and the violation of Bell's inequality, in which science solves a problem in its philosophy as well. Finally, I ask whether that ability justifies the inference that science is intrinsically epistemologically superior to its philosophy. At this point, I address the role that rhetoric plays in claims of epistemic superiority.

Closure in Science: The Example of Cold Fusion

Cold fusion has been a site of at least four book-length studies—all polemical. Although it is unclear that Eugene F. Mallove's *Fire from Ice: A Searching for the Truth Behind the Cold Fusion Furor* has a positive bias, the titles of the three negative polemics clearly signal their intent: Frank Close's *Too Hot to Handle: The Race for Cold Fusion*, John R. Huizenga's *Cold Fusion: The Fiasco of the Century*, and Gary Taubes, *Bad Science: The Short Life and Weird Times of Cold Fusion.* My interest lies not in defense or attack; I want simply to reconstruct the story of cold fusion historically, relying on archival materials at Cornell University and the University of Utah. It will turn out that to do so is at the same time to open this episode to rhetorical analysis.

The starting point for my historical reconstruction is a germinal article by historian of science Bruce Lewenstein. Lewenstein reflects on the speed and variety of communication in the controversy over the existence of cold fusion. He points especially to the negative effect of the rapid cycles of media and electronic communication on the stability that is a necessary condition for closure and for durable knowledge formation. His example is the English electrochemist David Williams, who "made a conscious decision to disregard information coming from outside the laboratory. His group felt they should focus on their own experiments, rather than trying to follow every twist and turn the others reported" (1995, 422). Lewenstein's point seems general. Some science at the cutting edge must turn into science on which a community can rely as a foundation for further research: there must be closure founded on the consensus of a scientific community.

As revealed in the audiotapes and transcribed e-mails from computer disks of the Cornell archive, cold fusion is a tale of consensus threatened and renewed, of trust tentatively given and firmly taken away. Consensus is the crucial variable, the basis of closure; in the general case, it consists of the set of facts, theories, and methods that are a discipline's intellectual capital. All disciplinary work assumes that this set is neither complete nor wholly correct; as a result, all disciplinary work is at least potentially disruptive, a threat to consensus. All disciplines, therefore, institutionalize a dilemma: unless the possibility of threat is preserved, there can be no new knowledge; unless the fact of consensus is protected, there can be no knowledge at all. The claim of cold fusion activated this dilemma in its most difficult form, across disciplines, and at its deepest level: the best theories of the physical universe predicted cold fusion's impossibility. Nevertheless, cold fusion was taken seriously as potential knowledge because Stanley Pons and Martin Fleischmann were trusted scientists; initially, they could be doubted only by provoking the self-doubt of the doubters themselves. As the controversy proceeded, there were two significant developments: first, Pons and Fleischmann acted in a way that eroded the provisional confidence of the audiences they addressed; second, as a result of social arrangements already in place, the locus of trust shifted: from Pons and Fleischmann to a procedure, peer review, and from peer review to a practice, experimental validation.

The story of cold fusion begins with a press conference in which two chemists, Pons and Fleischmann, pitted their professional reputations against the received wisdom of electrochemistry and nuclear physics. Prodded by confidence in their results, by insistent patent attorneys, and by ambitious university administrators, they chose to risk their considerable careers on the chance that they were right. It was a gamble: they were not experimentalists with the reputations of Albert A. Michelson or Robert Millikan, whose results had been accepted without question. But Pons was an established researcher, Fleischmann a Fellow of the Royal Society. Moreover, Pons and Fleischmann asked neither for implicit trust in their efforts nor for the routine confirmation of their results; they asked only to be proved wrong.

Given the importance of the result, the request, for concerned scientists who saw or read about the press conference, was virtually a mandate. Moreover, it was a mandate on a world scale. Although the press conference did not alter Pons and Fleischmann's message, it did what no professional publication could: in announcing their work to the world—literally and simultaneously, the world—they virtually guaranteed the immediate global attempt to replicate their work. If they were wrong, they would be wrong on a world scale.

If they were right, then two relatively obscure scientists would have achieved in five years of part-time work what "big science" working with expensive "hot fusion" had not achieved in two decades: the release of excess energy from nuclear fusion and, what is more, a release that would be cheap and easy. The discovery would make Pons and Fleischmann into world-class scientists, with the Nobel prize a virtual certainty; it would also make them rich. The discovery, moreover, would make Utah, its prosperity then faltering, prosperous indeed; it would also put the University of Utah, still smarting from the artificial-heart fiasco, on the world map. But the price of failure was equally high. If cold fusion failed, the university and the state would be laughingstocks, and Pons and Fleischmann disgraced, no longer members of *any* scientific community.

Given these stakes, it is little wonder that in interviews granted while events were unfolding and in a coincident stream of e-mail messages, we see a steady pressure on physicists and chemists to reach consensus concerning the fact or the fiction of cold fusion. This pressure begins immediately after the Pons and Fleischmann press conference of March 23, 1989, and rises exponentially thereafter. In the three days following the press conference, their lab received around four hundred phone calls. Science writer Eugene Mallove remembers that, at the Massachusetts Institute of Technology's Plasma Fusion Center, the professional community spent the evening of the twenty-third and the morning of the twenty-fourth intently watching Dan Rather and MacNeil-Lehrer reruns. Mallove sees this as a symptom of "an undercurrent of worry." This outsider's view is confirmed by an insider, Stanford University physicist Walter Meyerhof: "The hot-fusion people were quite upset with the whole thing" (Cornell Archive).

For this worry, James Brophy, Utah's vice president for research at the time, suggests two reasons: intellectual and financial. The hot-fusion community, he says, viewed the possibility of cold fusion as "an attack on their prerogatives." Cold fusion, a discovery of mere chemists, "challenged the well-known facts of . . . physics" (Utah archive). Moreover, a great deal of money was involved. In 1989, the annual operating budget for hot fusion exceeded half a billion dollars (Mallove 1991, 293–94; see also Herman 1990, 188–212). Cold fusion not only threatened these federal appropriations; the current costs of its investigation—an investigation mandated by the funding agencies, especially the Department of Energy—were also high. On August 22, 1989, a Los Alamos National Laboratory physicist "heard an estimate that cold fusion research is currently costing the research community something like $1 M[illion] a day, and that money has to be coming from somewhere. Most of it, undoubtedly, is being bootlegged from existing research grants." Most likely, the million dollars is an exaggeration, a sign, perhaps, of incipient panic. The remark about bootlegging, however, is amply confirmed throughout the interviews.

Only the degree of threat that cold fusion posed makes sense of the deeply personal nature of the attacks on its possibility. From the first, University of Utah physicist George Cassiday "personally concluded that it had to be nonsense . . . complete utter nonsense," but even he was taken aback by "the viciousness of the attacks" at the American Physical Society meeting in Baltimore six weeks after the press conference announcement: "Physicists can be real smart-asses . . . obnoxious and combative." This seems an apt description of one prominent physicist. A colleague, Charles Barnes, in a public lecture, said: "We are suffering from the incompetence and perhaps the delusions" of Pons and Fleischmann"; and at this point, "A hush fell over the hall" (Cornell Archive).

This view of Pons and Fleischmann's incompetence and delusion was extended to include the work of those who claimed replication. Much was made of the fact that these confirmations were coming from second-rate labs and third-rate countries (scientifically speaking, of course). California Institute of Technology chemist Nathan Lewis sneered, saying that results like Pons and Fleischmann are obtained "only if you have a football school." Brookhaven National Laboratory physicist George Jensen asked, "When is Tierra del Fuego going to announce they have done cold fusion?" Later, positive results from Stanford, Los Alamos, and Oak Ridge somewhat undermined these allegations; even Jensen himself came to believe that Pons and Fleischmann "have heat production, whether it's nuclear or chemical" (Cornell Archive).

In the attack at the American Physical Society meeting, Cal Tech's Albert Larson avoided "the F word," but accusations of fraud soon surfaced. University of California physicist Michael Nauenberg said, "I don't know where sloppiness ends and fraud begins, but certainly somebody's doing something that's wrong." Jensen confirmed this observation, "I know people who I trust . . . they've seen things that happened in Pons and Fleischmann's lab . . . that I can't understand unless there's fraud." Cal Tech physicist Harvey Lundt summarized these accusations, calling the work "a terrible blunder," a "fake." Indeed, fraud would seem the only justification for the heavy punishments that the accusing scientists thought were appropriate. Even the more moderate Jensen, also convinced of fraud, said, "If they're wrong, they should pay the price . . . they've destroyed Utah." At one point, Cal Tech's Larson told the interviewer, "I don't feel happy about it. . . . It's destroyed these two guys." When the interviewer added, "and their graduate student," Larson did not dissent (Cornell Archive).

The hostility to Pons and Fleischmann was not entirely gratuitous. Continuing trust depended on their openness, their willingness to share with the global scientific community even the smallest details of their experimental design. After all, scientists had been asked to envision an experiment with details gleaned from television news programs and newspaper accounts. Initially, this may have been understandable and forgivable, but Pons and Fleischmann's cold-fusion paper, circulated days later in preprint by fax and later published in the *Journal of Electroanalytical Chemistry*, was full of errors and empty of necessary detail. Moreover, subsequently published errata settled little. Worse, a second cold-fusion paper, submitted to the prestigious *Nature*, was withdrawn because, as Pons and Fleischmann improbably claimed, they didn't have time to provide the requested level of detail. As the controversy continued, Pons and Fleischmann became even more reclusive and less forthcoming.

In the end, the scientific audience's perception of Pons and Fleischmann's character only fueled but did not determine the rejection of cold-fusion claims. In the end, the locus of trust shifted from Pons and Fleischmann themselves to peer review and from peer review to experimental validation, a shift experienced within science as a move up the epistemological ladder toward increasing certainty and decisive closure.

Peer Review

There is no doubt that the press conference transformed peer review by involving the media and the general public in a process that ordinarily excludes these groups; there is no question that the depth of the controversy fueled existing biases and encouraged intense feeling. The real question is whether the involvement of the media and the general public biased peer review so seriously as to disrupt the building of an informed consensus concerning the existence and nature of cold fusion. California physicist Walter Meyerhof and Virginia Polytechnic Institute physicist R. Streater both told stories of unfair rejections, stories that might support the claim of disruptive bias. Meyerhof complained of irrelevance on the part of the *Science* referees: "One referee said . . . I quoted unpublished material from Huggins. . . . the other referee said he didn't find the results in his own lab." Streater complained that *Nuclear Physics B* gave him the runaround: "The fact is they don't want to publish a paper in favor of cold fusion" (Cornell Archive).

But such stories are routine among scientists. In fact, in the cold-fusion case, there is no substantial evidence that bias and intense feeling seriously influenced the professional judgments of the relevant scientific communities. The evidence points, rather, in the other direction, to the willingness of the scientific audience to give cold fusion a fair hearing. The six available sets of peer-review documents, admittedly, constitute a small and possibly biased sample considering the volume of submissions. If the number of papers on cold fusion submitted to *Nature* is typical, all of the journals, especially the most prestigious, were swamped. *Nature* received "about 800" manuscripts, according to David Lindley, an associate editor. Of the six available sets of reviews, four were from top-of-the-line journals, to which authors tend to send their papers first: *Nature* (2), *Physical Review Letters*, and *Science*; two were from journals of lesser prestige: *Nuclear Physics A* and *Fusion Technology*. Of the six papers submitted, three were accepted, three rejected. Of the three accepted papers, only one was critical of cold fusion; of the three rejected papers, this was also true. The case of the negative paper submitted to *Nature* is interesting for the light it sheds on referee fairness. The referee accepts the attack on cold fusion only on the condition that its authors eliminate entirely the mistaken implication that Pons and Fleischmann "were complete fools" (Cornell Archive).

So far as one can tell, Pons and Fleischmann themselves were well treated in the matter of their submission to *Nature*. According to Lindley, "Basically we wanted Pons and Fleischmann to do a much longer paper." The referee reports bear Lindley out. One asks for "a more complete description of the experiment . . . and more space . . . [devoted] to the presentation of raw data"; the other believes that "the MS deals with a topic of enormous interest to science and policy—all the more need to treat the data responsibly and present it fully to allow the reader to test its validity and implication." Both referees recommend acceptance, agreeing that a revision "should be published without delay." But Pons and Fleischmann withdrew their paper, refusing the opportunity afforded by peer review, that is, an opportunity to reestablish their authority by treating referees as collaborators rather than antagonists. Indeed, according to physicist Barnes, "They have not to this day [November 16, 1989] really made a detailed release of exactly what they did" (Cornell Archive).

Pons and Fleischmann were also generally well treated in the peer-review process; their decision to hold a press conference in lieu of peer review was not genuinely the object of criticism. True, chemist Kevin Wolf of Texas A & M University said that Pons and Fleischmann "started off on the wrong foot"; Michael Salamon of the University of Utah asserted, "Preliminary results should not be [put before] the press"; Cassiday judged that peer review by press conference "does science a lot of damage." The real objection, however, is not to the press conference itself but to Pons and Fleischmann's inappropriate and continuing reticence concerning the technical details of their work. MIT physicist Ronald Parker objected first to "holding press conferences," but then he put his finger on the real problem: Pons and Fleischmann are to be blamed for "not telling details." According to Cal Tech's Larson, "Pons and Fleischmann should have allowed their paper to be refereed; they submitted it to *Nature*" (Cornell Archive). Larson is wrong on the matter of fact. The paper was refereed by *Nature*, but his point about openness stands.

The evidence in the cold-fusion controversy supports a verdict that peer review remains a trusted and trustworthy resource for determining worth within scientific communities. But only preliminary worth is at stake. In peer review, referee judgments merely *represent* the judgment of the disciplinary community; they do not actually *embody* that judgment. In any matter of importance, moreover, the gap between mere representation and actual embodiment is likely to be telling. At that level, peer-review consensus will be no substitute for the actual consensus of the community. In addition, the evidence supports a verdict that peer review cannot free itself from *allegations* of bias, allegations every scientist can support with personal testimony. In the case of cold fusion, Cassiday complained of "old boy networks" and of a preference for "fashionable" projects. Meyerhof said, "The believers are having a hard time getting their material published." Streater spoke of "definitely hostile referees" (Cornell Archive). As sociologists Daryl E. Chubin and Edward J. Hackett make clear in their analysis of federal peer review, such allegations are typical. Because peer review does not reflect actual community consensus and cannot free itself from allegations of bias, allegations for which virtually every scientist can offer testimony, its judgments cannot seriously contribute to the resolution of a controversy so deeply divisive as cold fusion.

Experimental Disconfirmation

There was an additional roadblock to successful resolution: the very real, very intense rivalry between two disciplinary communities, the chemists and the physicists, a rivalry that the interdisciplinary nature of cold fusion brought to the fore. Chemist Jack Simons alluded to a deluge of e-mail from physicists offering unsolicited "opinions on why [cold fusion] is probably wrong"; chemist Wolf referred to "the lynching APS [American Physical Society] meeting in Baltimore." The motive for this intemperate language can be overheard in the condescension of physicist Jensen when he said of chemists, "They've grown up through a more alchemical approach [but] they're making great strides." The resentment of chemists and the condescension of physicists were fueled by status differences endorsed and enforced by federal funding agencies: "Chemistry in particular has a very bad image even in that subset called science"; as a result, "electrochemistry hasn't been a heavily funded field" from federal sources, Jensen said (Cornell Archive).

This rivalry made intellectual accommodation impossible. At first glance, it would seem possible to build theoretical bridges between what were, after all, two physical sciences that shared the same microworld. In pursuit of this goal, chemist Simons said, "I quickly became a nuclear physics student." Of the gap between nuclear physics and chemistry, he opines:

> It sounds sort of crazy but in all honesty the theory of processes and structures of nuclei are fairly similar to electronic structures of atoms, in other words . . . nucleons occupy orbitals like electrons do; the thing that's different is that in an atom you have a plus charge in the center with the electrons moving around it; and in the nuclear shell model, it's called, you instead have this spherical volume element that is the nuclear volume in which the protons and neutrons

> move without a center of force that's like the plus charge . . . orbital structure is somewhat similar. (Cornell Archive)

There were profound differences in scientific practice between two very different disciplinary communities, the electrochemists and the nuclear physicists, differences that precluded the building of intellectual bridges. "Each science," physicist Nauenberg said, "has its own language." This may be taken literally, of course. A referee commented on a cold-fusion paper by physicists: "If the authors use [this] terminology, . . . they will cause great troubles among electrochemists. Thus, Θ is used by electrochemists always to denote the fraction of the surface cover to be adsorbed by hydrogen or other active radical." Nauenberg's statement also bears a figurative interpretation, one not so easily addressed. According to Los Alamos physicist Martin Clay, "Chemists working on cold-fusion research appear to have generally different criteria for judging 'confirmation' than do physicists" (Cornell Archive). The reason may be their differing working ontologies, an artifact of differing methodological practices. Nuclear physicists think in terms of subatomic particles, the building blocks that make up atoms; electrochemists think in terms of the atoms themselves, the building blocks that make up molecules.

Whatever their cause, these differences could, as Jensen put it, be "very frustrating," a frustration evident in the way the members of each discipline treated the results of the other: not as authentic science but as a consequence of human failings. To those on Pons and Fleischmann's side, such as Brophy, "The response of the physics . . . fusion community is a study . . . in the sociology of science"; to those on the other side, such as Larson, cold-fusion research "ain't science, it's a mix of politics, sociology, psychology" (Cornell Archive). Given these deeply held views, consensus through intellectual accommodation proved impossible.

Nevertheless, consensus was achieved through a resource shared by both communities; each agreed that the existence of cold fusion was defined by a method, "the scientific method," whose central feature is the verification of experimental data through replication. The mythical character of this method may be conceded (Chalmers; Schuster; Collins) but only if the referent is to some general method for ensuring valid intellectual results, such as the Baconian or the Cartesian. In this case, however, the referent is rather the sum of the methods these scientists and their colleagues employ, the confluence of a set of actual practices.

Such practices are a source both of "eureka" anecdotes of discovery and of refutational accounts. Wolf's group was ready to abandon its attempt when members observed experimental data they could not gainsay: "[We saw neutrons] 9 deviations or 10 above background and then we were hooked." Opponents also referred to the recalcitrance of experimental data when they undercut cold-fusion claims: Charles Martin of Texas A & M stated that his "belief in the excess heat effect began to crumble as we did more and more experiments" (Cornell Archive).

In each case, replication is the key, the stability of the data sets over time and from experimenter to experimenter. "You can never, I think, disprove an experiment, you'd better not do that, OK?" said Nauenberg. "An experiment is an experiment." Although it cannot be "disproved," an experiment can be dismissed and its results ignored if its replication fails. "Replication . . . sure, it's the scientific method," said Wolf. Ohio physicist Edward Kaufmann said:

> I'm a nuclear physicist, but nothing in my reply requires expertise in nuclear physics. Barring alchemy, if one gets more tritium out of an electronic cell than is put in, fusion has occurred. The trick is to conclusively demonstrate that more tritium is found in the electrolyte (or the electrode) than could have been introduced by ANY source in the cell!

But, in the case of cold fusion, "There is a question: 'Is there any experiment which has been carefully performed . . . which shows a positive result?' . . . The best and most careful experiments with the most controls (e. g., Frejus [the France-Italy tunnel], Yale BNL [Brookhaven National Laboratory]) find nothing" (Cornell Archive).

Martin emphasized the central concern: the "result was very hard to reproduce." It was this difficulty that was perceived by *both* physicists and chemists as sufficient reason eventually to abandon the project. Physicist Dennis Manos indicated that attempts to replicate might have been different and differently received "if [Pons and Fleischmann] made a claim, published a recipe." Electrochemist Williams justified the testing shutdown at Harwell: "If you can give us a good recipe and convince us it's a good recipe, then we'll do it again" (Cornell Archive).

In the case of cold fusion, closure had been achieved by means of experimental disconfirmation. That science has the resources to solve its own problems, even in a case so deeply divisive as cold fusion, is impressive. More impressive yet would be its ability to solve problems not only in its own domain but in the domain of its philosophy as well. From this ability, indeed, it might be possible to infer that science possesses an *intrinsic* epistemological superiority. Apparently, this is the lesson to be learned from the solution of a particularly recalcitrant puzzle, the nature of quantum reality.

Closure in Philosophy: Philosophy of Physics

In the early twentieth century, in a series of exchanges, Albert Einstein and Niels Bohr argued over quantum physics. They differed not concerning the predictive power of quantum theory but concerning its status as an accurate description of an underlying reality. Bohr felt that whatever quantum physics said about the microworld was more than a convenient fiction that licensed accurate prediction; it was also true of that world. Einstein thought otherwise; he felt that this new science was simply a successful formalism, a way station to a more complete physics, one that would reinstate the robust objectivity characteristic of classical physics. Bohr and Einstein were arguing philosophy, not physical theory; at the time, the theoretical and experimental resolution of the problem they confronted were out of the question.

Einstein challenged the assumptions of quantum physics concerning the nature of reality. He did so by means of a series of "thought" experiments, a series of science fictions created in the interest of science fact. Because the world of thought experiments is a world whose causally ordered constituents scientists share, their fictional status does not prevent them from bearing importantly on real science. Thought experiments are designed to reveal unexpected implications of that shared world, inferences at odds with received wisdom concerning its constitution. So it was with the most famous of Einstein's thought experiments, the EPR experiment of 1935, named for its coauthors, Einstein, Boris Podolsky, and Nathan Rosen.

Quantum mechanics is the physics of the microworld: the atom and its constituents. Among physicists, there is general agreement that this microworld obeys Heisenberg's uncertainty principle: if we measure the momentum of an elementary particle such as the electron, we cannot determine without uncertainty the position it held at the time that measurement was made. This limit is built into the microworld because to measure any of the parameters of this tiny universe is so to disturb its state that our ordinary assumption of an objectivity independent of our measurements, the standing assumption in our macroworld, must be suspended.

EPR asks us to imagine a possible world in which Heisenberg's principle holds. In this world, there are two particles in close proximity, each with a position and a momentum. Let us call them Mary and John. Although Heisenberg's principle does not permit us to measure the position and momentum of Mary or of John at a single time without uncertainty, it does permit us to sum the momenta of these two particles and to measure the distance between them. We do this. John and Mary then interact, and Mary flies far off, a distance so considerable that interaction between the two particles seems impossible.

So far, the EPR thought experiment has respected Heisenberg's principle; now, in apparent violation of that principle, EPR allows us to fix the simultaneous position and momentum of Mary. The procedure is simple. To fix the position of Mary, we subtract the original distance between the two particles from the new distance, figures obtained without in any way disturbing the system. Then we measure the momentum of John and subtract it from the total known momentum and, as a result, obtain the momentum of Mary. This measurement, to be sure, makes it impossible to determine the position of John without uncertainty, but it cannot disturb the position of Mary, far away.

The Heisenberg principle, however, states unequivocally that the momentum and position of Mary at a single time cannot be determined without uncertainty. Therefore, quantum mechanics contains a central inconsistency: according to its laws, the position and momentum of Mary at a single time *can* and *cannot* be determined without uncertainty. To remove the contradiction, EPR maintained, we must suspend local causality, the assumption that allows its measurements to be made, the assumption that the behavior of John can have no influence on that of far-away Mary. But to suspend local causality would be to move from EPR to ESP, a move few physicists would make. Thus, quantum mechanics must be incomplete; it cannot be the final word in physics.

Bohr disagreed with the EPR conclusion that quantum mechanics contained a contradiction; he insisted rather that EPR was guilty of a serious equivocation in its definition of thinking physically. Quantum mechanics demonstrates that we must think physically in a new way. It teaches that the observer is part of the system; to think physically about the microworld, therefore, we must include the influence of the observer in our calculations. Whatever the Heisenberg principle entails, whatever observer dependence entails, we must learn to live with it. It is not quantum mechanics that is incomplete but EPR's view of what it is to think physically.

Physics Solves a Problem in Philosophy

The philosophical differences between these two physicists were settled not by more philosophy but by more physics; these differences were decided nearly half a century later by a theoretical formulation called Bell's inequality and a series of experiments demonstrating that the elementary particles of the microworld consistently violated that inequality. In this reconceptualization, the Einstein-Bohr philosophical dilemma proved amenable to experimental resolution.

To understand Bell's inequality, the theoretical formulation that finally broke the philosophical deadlock between Bohr and Einstein, we will make do with a rough analogy: experiments actually made with light and polarization will be envisioned as experiments in which two toy guns are back to back, each aimed outward at a slotted target. The target slots, A and B, are set at the same angle. At regular intervals, the two guns fire simultaneously toward their respective targets pairs of projectiles shaped like blackboard erasers. Each projectile of a pair matches the orientation of the other, but the orientation from pair to pair is random. Only two outcomes are possible: either the two projectiles pass through the slot or fail to do so. There are two hits, or there are two misses:

A: 001011001
B: 001011001

Let us now alter the angle of slot A by turning it 15° in a clockwise direction. If we do, a third outcome becomes possible: a hit on one side coupled with a miss on the other. This generates an error rate from B's point of view. When we repeat this procedure, turning B 15° counterclockwise, we generate an error rate from A's point of view. Because both alterations in angle were the same, the error rate for A and B together is exactly double the error rate for A or B taken separately. Let us add these two sets of errors to obtain their sum, E_1. For example:

A: 0010011010
B: 0000101010

In this double sequence, we have three mismatches, or errors, an error rate of 30%.

As a final step, let us simultaneously alter the angle of *both* slots. This simultaneous alteration creates a new situation, one without a standard. In this new circumstance, a new sort of error becomes possible, a double error. Suppose we have a miss at A; we register a 0. But because we have shifted B also, we can also have a miss at B: two possible hits, a 1 and 1, have become actual two misses, a 0 and a 0. Because double errors do not show up as mismatches, they will be undetectable.

Continue this imaginary experiment according to this last condition, firing the two guns again at regular intervals. This last trial will also yield an error rate E_2. Because of double errors, in the aggregate, E_2 will always be less than E_1. This is Bell's inequality.

In the world of middle-sized objects, our world, Bell's inequality invariably holds; in the microworld of quantum mechanics, it never does. If, in the Bell experiments, photons existed in an objective state, if they existed not dispositionally or potentially but as palpable objects, then the result at one target slot must have influenced the result at the other; otherwise, the Bell inequality would have held. But influenced how? By some sort of ESP? On the other hand, if there is no ESP but only the down-to-earth local causality to which physicists—and most ordinary folk as well—subscribe, then photons cannot have existed in an objective state. Thus, the invariable microworld violation of the Bell inequality forces on physicists a choice not forced by the macroworld, the counterintuitive choice between local causality and objectivity.

> In the original EPR paper, Einstein and his co-authors insist that any serious consideration of a physical theory must take into account the distinction between the objective reality, which is independent of any theory, and the physical concepts with which the theory operates. These concepts are intended to correspond with the objective reality,

> and by means of these concepts we picture this reality to ourselves. (1935, 777)

In the light of Bell's inequality, this argument collapses. The objectivity of the microworld is transformed from an assumption to a contention, one that may not hold; a philosophical debate had been transformed into a theoretical problem, a problem that was to be solved experimentally.

The experimental confirmation was ingeniously accomplished a half-century later by Alain Aspect and his collaborators. In Aspect's laboratory, the EPR thought experiment became a real experiment at last: as a consequence of Aspect's arrangements, real photons obeyed the predictions of quantum theory while their behavior violated the Bell inequality by *five standard deviations*:

> Two runs have been performed in order to test Bell's inequalities. In each run, we have chosen a set of orientations leading to the greatest predicted conflict between quantum mechanics and Bell's inequalities
>
> $$[(\vec{a},\vec{b}) = (\vec{b},\vec{a}') = (\vec{a}',\vec{b}') = 22.5°; (\vec{a},\vec{b}') = 67.5°]$$
>
> The average run yields
>
> $$S_{expt} = 0.101 \pm 0.020$$
>
> violating the inequality $S \leq 0$ by 5 standard deviations. On the other hand, for our solid angles and polarizer efficiencies, quantum mechanics predicts $S_{QM} = 0.112$.
>
> We have carried out another run with different orientations, for a direct comparison with quantum mechanics. Figure [9.1] shows that the agreement is excellent. (1982b, 1806)

Realistic locality theories, such as that of Einstein's, had been tested against experience and found wanting. In the words of Fritz Rohlich in *Science* in 1983, "Local hidden variables theory is dead. It received its coup de grace by two precision experiments carried out last year in Paris" (1251). For technical reasons, Aspect's experiments testing Bell's inequality were not "a fully conclusive argument against the whole class of supplementary-parameter theories obeying Einstein's causality"; nevertheless, "the overwhelming majority of physicists who are actively engaged in fundamental research involving QM on the atomic and subatomic level" accepted this result as decisive (Aspect, Grangier, and Roger 1982b, 1807; see also Aspect, Grangier, and Roger 1982a; Rohlich 1983, 1251; Ballentine and Jarrett 1987). In many quarters, this was a source of joy: "At a recent public lecture, Harvard physicist and University Professor Emeritus Edward Purcell told

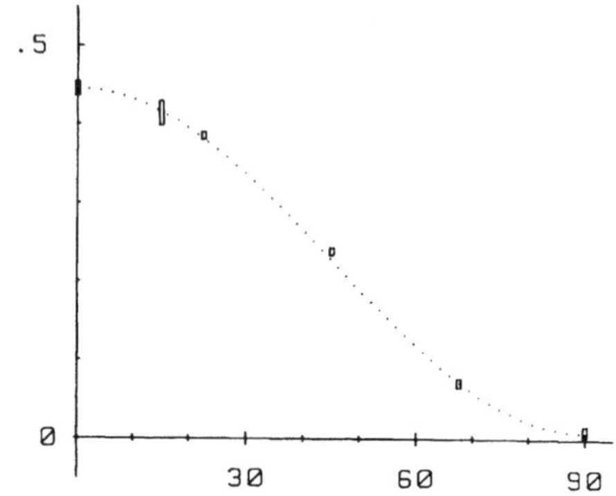

Fig. 9.1. The Death-Knell of Quantum Realism. Average normalized coincidence rate as a function of the relative orientation of the polarizers. Indicated errors are ± one standard deviation. The dashed curve is not a fit to the data but the predictions by quantum mechanics for the actual experiment. Reprinted figure from A. Aspect, P. Grangier, and G. Roger, "Experimental Test of Bell's Inequalities Using Time-Varying Analyzers," *Physical Review Letters* 49 (1982): 1804–7. Permission of Prof. Alain Aspect and the American Physical Society. Copyright © 1982 by the American Physical Society.

his audience of his happiness that he has lived to see a philosophical problem settled in the laboratory" (Jarrett 1989, 60).

The Physical Chicken or the Philosophical Egg?

The solution of these scientific and philosophical problems by means of experiment suggests that epistemological ranking, a ranking that privileges science over philosophy as a way of knowing, is a notion far sturdier than Rorty claims. Nevertheless, we must not draw from these cases the more radical conclusion that the ability of experimental science to solve philosophical problems confers on science an intrinsic epistemological superiority, an authority independent of any arguments on its behalf. The experimental epistemology of modern science, an epistemology on which the closures of cold fusion and the quantum violation of Bell's inequality depend, were argued into place in the seventeenth century, most notably by Robert Boyle, Francis Bacon, and Isaac Newton, and the triumph was as much rhetorical as it was scientific.

In *The Origin of Forms and Qualities* published in 1661, but written a decade earlier, Boyle argues for the plausibility of that central ontological presupposition of early modern experimental science, the machine-like nature of the microworld, an essence prerequisite to its law-like description: "That . . . which I chiefly aim at is to make probable to you by experiments . . . that almost all sorts of qualities . . . may be produced mechanically—I mean by such corporeal agents as do not appear either to work otherwise than by virtue of the motion, size, figure, and contrivance, of their own parts" (1979, 17).

To Boyle, the marvelous variety that constitutes the macroworld suggests an analogous variety in the constitution of the microworld: "The shape of vitriol [any of a family of crystalline salts] depends upon the textures of the bodies whereof it is composed" (1979, 86). To confirm this point, he precipitates crystalline shapes that vary according to the substance dissolved:

> Besides the lovely vitriol of copper formerly mentioned, that liquor would with quicksilver afford one sort of crystals, with silver another, and with lead a third: all which crystals of vitriol, as they differed from each other in other qualities . . . so they did very manifestly and considerably differ in shape, the crystals of silver shooting in exceeding thin plates, and those of lead and quicksilver obtaining figures, though differing enough from each other, yet of a far greater depth and thickness, and less remote from the figures of common vitriol or seasalt. (1979, 86)

But there is for Boyle a more impressive demonstration of the essentially mechanical character of the microworld: nature's ability to reconstitute even its more complex forms after their apparent obliteration as a result of dissolution or chemical reaction. What better explanation of this reconstitutive power than a memory for mechanical arrangements, analogous to human memory? To confirm this point, Boyle reconstitutes antimony, revealing

> what some [scientists] would call the mindfulness of nature: which, when a body was deprived of a not inconsiderable portion of its chief ingredient, and had all its other parts dissipated and shuffled and discoloured, so as not to be knowable, was able to rally those scattered and disguised parts, and marshall or dispose them into a body of the former consistence, colour, & co., though (which is not here to be overlooked) the contexture of antimony, by reason of the copious shining *stirae* [streaks] that enoble the darker body, be much more elaborate, and therefore more uneasy to be restored, than that of other concretes. (1979, 95–96)

Boyle's mechanical ontology is complemented by an experimental epistemology, one that presupposes that practices confined to the laboratory nevertheless yield information about actual processes in nature. Boyle's argument for this epistemology is as analogical as its ontological counterpart, equally a comparison between nature and a machine: "If art can produce vitriol itself as well as nature, I see not why we may not draw arguments from the qualities and operations of several of those [bodies] that are called factitious [artificial], to show how much may be ascribed to, and performed by, the mechanical characterization or stamp of matter" (1979, 86–87).

Boyle's epistemology rests on Baconian induction, the climax of which is "the crucial experiment," a test of nature decisive between theories. Bacon's example involves two clocks, one moved by lead weights, the other by an iron spring. By placing the clocks at various heights, we can measure the force of gravity and determine its cause. If the spring-driven clock runs the same no matter what its distance from the center of the Earth while the weight-driven clock varies depending on this distance—it runs faster the closer it is to the center—we are measuring the force of gravity. We can also determine the cause of this force: "If the virtue [efficacy] of the weights is found to be diminished on the steeple and increased in the mine, we may take the attraction of the mass of the earth as the cause of the weight" (Bacon, *Novum Organum* 2:xxxvi).

It is Newton who transforms this rhetorical invention into a rigorous practice. The climax of his first paper on optics is precisely an "*Experimentum Crucis*," alleged to be decisive between theories (1978, 50). In his optical works, moreover, Newton establishes experimentation as the sole arbiter among competing explanations of natural phenomena; experiments decide not whether the universe is fundamentally machine-like but which version of a machine-like universe is correct. In the light of current knowledge, Boyle's naively strict analogies between nature and machines need to be revised to include fields. In addition, although the experiments of Aspect come close to the ideal, the youthful Newton's naive reliance on the persuasiveness of the crucial experiment needs to be seriously modified. These revisions, however, do not affect my argument: the experimental empiricism criterial of contemporary science was argued into place in the seventeenth century.

That science solved a philosophical problem, therefore, gives it no undue epistemological credit. The problem was "solved" only by a reconceptualization that made experimental methods decisive, but it was only by arguments from analogy that this reconceptualization occurred. The deflationary arrow points in the other direction as well. That philosophy has access to knowledge unavailable to natural science, namely, the knowledge that makes natural science possible, cannot elevate philosophy above science. Such claims systematically confuse the possibility with the fact of science: philosophy is neither a sufficient nor a necessary condition for the existence of the sciences.

The lesson to be learned from these cases is not that science is rhetoric; it is that no one discipline—neither rhetoric nor history nor philosophy nor sociology—can be relied on to provide a fully adequate explanation of what science is and how it works. The final two chapters of this book address this interdisciplinary issue. Before I address it, however, in the next chapter, I turn to a final topic within the range of rhetorical analysis—the relation of science and society.

10 Science and Society

Curing Occupational Medicine

Western medicine is a realization of the first of Jürgen Habermas's three cognitive interests, that of technical control: human beings are seen as living machines. This view is embodied in the biomedical model: health is the unimpaired functioning of these machines; disease, a causally linked malfunction amenable to treatment. Simultaneously, Western medicine is a realization of Habermas's second cognitive interest, the practical (1971, 309–10). Physicians form a society within a society, dedicated to thc full institutionalization of the biomedical model, and create a system that fully governs the treatment of disease by embodying that model in a network of social, political, and economic purposes. When this confluence of interests occurs, a moral order emerges, a synergy of structures and values so finely tuned and smoothly functioning that, although clearly historical in origin, it seems the only arrangement possible, one virtually coincident with the natural order. But moral orders differ: they are "a way in which ruling groups create legitimation and functional response to their power and interests." They do so "by construction of a cognitive and moral reality, a set of motives and directions in the ruled which are consonant with the needs and interests of ruling groups" (Gusfield 1981, 187).

Because occupational medicine privileges the biomedical over other models of disease, it easily coincides industrial purposes; it readily views human beings not only as machines but as human capital: "The positivist self-understanding of the nomological [law-governed] sciences lends countenance to the substitution of technology for enlightened action" (Habermas 1971, 317). Nevertheless, despite the social, political, and economic forces arrayed against them, workers sometimes engage in successful enlightened action; sometimes, they realize Habermas's third cognitive interest, the emancipatory, an interest that "releases the subject from the hypostatized powers," such as those of medical science allied to industrial regimes (Habermas, 1971, 310).

To accomplish this emancipatory goal, workers must turn social stasis into social drama; they must remake a moral order that has forged a coincidence of purpose between the biomedical model of disease and industrial imperatives. They must create a new moral order based on a different model of disease, the ecological, in which the environment itself can count as a pathogen. They must do so by creating a breach in the existing moral order, and, by means of a rhetoric of talk coupled with a rhetoric of action, they must redefine human health in a way that serves their interests.

Occupational Medicine as a Moral Order

As old as manual work, occupational disease has been noted in the medical literature since antiquity; as early as 1700, it was described in comprehensive fashion by Bernardino Ramazzini. None of this early work is scientific in our sense; none constitutes occupational medicine as a discipline. For that to happen, its founding science, epidemiology, had to be established: without a knowledge of the causes of wide-spread diseases, occupational disease, a wide-spread malaise caused by adverse working conditions, could neither be diagnosed nor cured. According to Richard R. Monson, author of a standard textbook in the field, the groundbreaking effort at creating the discipline of epidemiology was that of John Snow, a mid-nineteenth-century London physician with an interest in the spread of cholera. Snow is best known for his role in the removal of the handle of the Broad Street pump, whose water was contaminated by the cholera bacillus. Here is Snow's account:

> As soon as I became acquainted with the situation and the extent of this irruption of cholera, I suspected some contamination of the water of the much-frequented street-pump in Broad Street. . . . On proceeding to the spot, I found that nearly all the deaths had taken place within a short distance of the pump. . . . I had an interview with the Board of Guardians of St. James's parish, on the evening of Thursday, 7th September, and represented the above circumstances to them. In consequence of what I said, the handle of the pump was removed on the following day. (Snow 1936, 38–40)

Monson regards the ameliorative effect of these events as a prelude to the crucial step in disciplinary foundation, the provision of a scientific model: "John Snow's contribution to the history of epidemiology lies in the *next* steps he took" (1990, 6; my emphasis). In one London region supplied by two competing water companies, Snow recognized that he had a natural laboratory: "It is obvious that no experiment could have been devised which would more thoroughly test the effect of water supply on the progress of cholera" (1936, 75). This language is neither socially nor politically innocent. It represents a metonymic shift: science contributes to human flourishing by viewing human beings as experimental subjects. For Monson, the crucial factor in disciplinary formation is not that Snow's work led to the prevention of cholera but that prevention was possible only through medicine conceived of along the lines of an experimental science. Lives were saved by means of good experimental design: isolation of the independent variable, the water already identified with the spread of cholera, and a satisfactory control.

The moral order that is occupational medicine depends for its high status on biological science, especially the definition of disease in terms of a definite pathology, a dependent variable causally linked to an independent variable in the workplace. Its basis is a metonymic shift from workplace conditions generally to a causal chain capable of isolation; its basis is the integration of everyday medical practice and the experimental norms of the laboratory sciences (Foucault 1972, 52). In *Birth of the Clinic*, Michel Foucault identifies a central consequence of this phenomenon: the transformation of disease into an entity that must be defined by means of specifying "the point from which [its] pathological organization radiates" (1975, 140). From this definition stems a classification of disease based on invariable organic anomalies, a classification in which the "symptoms speak the very language of pathological anatomy" (1975, 129, 187). On this issue, Foucault's views conform to more orthodox opinion. Writing in *Science*, George L. Engel characterizes contemporary medical diagnosis: "Taxonomy progresses from symptoms, to clusters of symptoms, to syndromes, and finally to diseases with specific pathogenesis and pathology. This sequence accurately describes the successful application of the scientific method" (1977, 131). This is a view of disease that conforms exactly to the criteria of Habermas's first cognitive interest, the technical.

For the development of epidemiology, a second methodological ingredient was necessary: statistics. Epidemiology usually deals with large populations, often studied in retrospect. In these cases, causal links must be discovered by establishing correlations between specific environmental factors and specific pathological conditions. Vital statistics and simple correlations were a part—though a small part—of medicine from the mid-nineteenth century on. For example, Ignaz Semmelweis linked sepsis with puerperal fever around mid-century; toward the century's end, Rudolf Virchow used vital statistics in his public-health studies, for example, in "Typhus and Municipal Sanitation: "Whoever looks at the curve in [Herr Albu's] Table I without being convinced that in Berlin a close correlation exists between typhus and groundwater is beyond help. This table deals with the years 1870–1873, and gives for *each of these years* a 'striking proof' that typhus mortality increases with a decrease in groundwater level, and vice-versa" (1985, 418). But the real transformation in epidemiology occurred with the shift from vital to mathematical statistics—the whole now-familiar apparatus of chi-squares, Student *t*'s, and Pearson product-moment correlation coefficients—a development that took place in the early decades of the twentieth century due to the influence of Karl Pearson and his students (Magnello).

Occupational Medicine Delayed

Between the discovery of an appropriate scientific model and the constitution of a new medical specialty of occupational medicine, a considerable delay intervened. Monson does not attempt to account for this delay. He merely notes that "between the middle of the 19th century and the middle of the 20th, epidemiology progressed slowly" (1990, 8) and that, as a consequence, occupational medicine was slow to form as a discipline, a society within a society that realized Habermas's second cognitive interest, the practical. What accounts for this delay? The answer lies in the initial lack of coincidence between the interests of occupational medicine and those of industry. Concerning the need for such a medical specialty, there can be no question. Available data document the health hazards of the workplace and the poor health of workers from Ramazzini to twentieth-century America. In 1927, for example, the *Journal of the American Medical Association* published a study of thirty-nine American enameling workers, twenty-three of whom showed evidence of lead poisoning. Of the thirty-nine, there was "not a single individual who could be given an absolutely 'clean bill of health'" (Leathers and Morgan 1109). Most surprising in a population whose average age was twenty-seven, seven men had arteriosclerosis, and three suffered from organic heart disease. The relative youth and poor health of those in the dangerous trades in America is documented as well in a 1928 study of 267 hatters: "Boys 20 to 21 years old are already so badly poisoned that their hands shake continually, while many of the men who have served longer at the trade cannot even feed themselves" (qtd. in Rosner and Markowitz 1989, "Safety" 59).

As a consequence of earning a living in this century, American workers in the dangerous trades were dying at alarming rates. In 1915–1918, the number of deaths of Vermont granite cutters from pulmonary tuberculosis was nearly ten times the state average (Winslow 1925, 970). In 1909–1919, in a Connecticut ax factory, the pulmonary tuberculosis death rate of grinders and polishers was eleven times the state average of employed males. In each industrial case, the cause of death was the inhalation of silica dust, the cause also in the case of 467 workers building a tunnel for Union Carbide between 1931 and 1933 (Rosner and Markowitz 1989, introduction, xvi–xvii).

Industrial neglect was complemented by the evident powerlessness or indifference of unions in matters of worker health. All 267 hatters, poisoned by their work, were union members. Unprotected by local membership, they were unprotected also by national affiliation. In 1890, the feuding Knights of Labor and National Federation of Miners settled their quarrels and formed the United Mine Workers. However, this concord had no apparent effect on worker health and safety: deaths in the coal mines remained constant between 1875 and 1900 at about three per thousand (Corn 1989, 77, 80), while between 1907 and 1912, accidental deaths among coal miners were two and one half times the national average (Bale 1989, 37). The Workers' Health Bureau, the union agency that conducted the study and pressed for reform, was active for only eight years, from 1921 to 1928. Though supported by union locals throughout the country, it disappeared because of lack of support at the national level (Rosner and Markowitz 1989, "Safety" 53–63).

Union powerlessness or indifference was complemented by government ineffectuality in protecting worker health and safety. A Pennsylvania mine disaster in 1869 in which 179 miners suffocated led to state legislation that protected anthracite but not bituminous miners; the bituminous miners did not have similar protection until 1877. Ohio and Illinois also passed mine-safety bills in the 1870s but not Indiana or West Virginia, the other leading coal-mining states. Although by the end of the century, twenty-one mining states passed safety laws, these were essentially ineffective. In 1900, there were 1,489 known deaths, 5.12 deaths for every thousand short tons, 3.44 deaths for every thousand miners (Corn 1989, 78–79).

At the federal level, help was also slow in coming: the Federal Coal Mine Health and Safety Act was not passed until 1969. Moreover, there persists to the present day a problem of coordination and control: a de facto division between the roles of the Department of Labor and the Public Health Service, agreed to during the New Deal, is concretized in current legislation. Under this legislation, the Public Health Service provides technical expertise, the Department of Labor the policing function. This division of functions is virtually a formula for inefficient enforcement (Rosner and Markowitz 1989, "Research" 83–99).

So long as the immigrant tide flowed, all trades—and in particular the dangerous trades in particular—of mining, asbestos, textiles, and paint manufacture—could be readily resupplied from a working class protected only by ineffective state and federal governments and by unions indifferent or powerless to protect worker health. This flow is reflected in high worker turnover in the dangerous trades: at the time, the average age of enamel workers was 27.1 and the average time of employment thirteen months; in addition, 98% of the hatters examined by the Workers' Health Bureau were between twenty-five and forty-five years old (Leathers and Morgan; Rosner and Markowitz 1989, "Safety" 57).

From 1870 to 1930, a policy of open immigration favored industry by creating a steady supply of labor. From 1890 to 1924, for example, over twenty million immigrants accounted for nearly 35% of the population increase. By 1909, in twenty-one leading industries, nearly 60% were immigrants. So long as a resupply of cheap, docile labor was secure, and union and government neglect guaranteed, it was in industry's interest to ignore, not to explore the connections between disease and the workplace.

Occupational Medicine Emerges

It was not until the pool of immigrant labor had diminished markedly that industry perceived a need for medical science related to worker health. It is not coincidental that the years between 1930 and 1959, a time when the American Association of Industrial Physicians and Surgeons surged in membership from a 315 to 4,000, is also a time when the number of new immigrants and the cheap labor they supplied plummeted. Although after 1930, occupational disease was acknowledged in the general case, industry labored to suppress knowledge of its specific instances, to curb or sidetrack research unfavorable to these interests, and to encourage studies that directly or indirectly favored corporate purposes.

In the crassest of strategies, company physicians simply colluded with industry by systematically misinforming workers concerning their state of health. In 1933, for example, Union Carbide company doctors systematically misled tunnel drillers concerning the mortal dangers of silicosis (Rosner and Markowitz 1989, introduction, xvi–xvii). Company physicians at Johns Manville behaved similarly concerning the threat of asbestosis. In 1949, Dr. Kenneth W. Smith, corporate medical director, expressed the corporate medical policy in the following confidential memorandum:

> It must be remembered that although these men have the X-ray evidence of asbestosis, they are working today and definitely are not disabled from asbestosis. They have not been told of this diagnosis for it is felt that as long as the man feels well, is happy at home and at work, and his physical condition remains good, nothing should be said. When he becomes disabled or sick, then the diagnosis should be made and the claim submitted by the Company. The fibrosis of this disease is irreversible and permanent so that eventually compensation will be paid to each of these men. But as long as the man is not disabled it is felt that he should not be told of his condition so that he can live and work in peace and the Company can benefit by his many years of experience. Should the man be told of his condition today there is a very definite possibility that he would become

> mentally and physically ill, simply through the knowledge that he has asbestosis. (qtd. in Kotelchuck 1989, 203)

Unfavorable medical research on occupationally related diseases was also curbed or suppressed. The career of Dr. Leroy Gardner illustrates this second strategy. In a typical incident, Gardner was consulted about an outbreak of a lung disease at the Salem, Massachusetts, plant of Sylvania. The possibility existed that the disease was linked to the beryllium used in the manufacture of fluorescent lamps. Two years after the initial outbreak, action had still not been taken. At this point, Gardner was asked by a company representative to review a paper on workplace conditions and beryllium disease at the plant. The representative was disturbed that the paper "gives an indication that this disease may be of occupational origin" (qtd. in Zwerling 1989, 111). Before the company intervened, Gardner had enthusiastically recommended publication; now, in deference to this concern, he suggested that this indication might be omitted. Later, a compromise was discussed, in which the question of occupational origin would be raised rather than asserted. In the end, the paper was never published (Zwerling 1989, 103–15). It should not assumed that the 1970 passage of the Occupational Safety and Health Act eliminated this industrial strategy. A case of employer duplicity surfaced as late as 1977, seven years after OSHA. Research had been published as early as 1961 showing that exposure to DBCP (dibromochloropropane) caused sterility. Despite this knowledge, workers at American Cyanmide who were routinely exposed to this chemical were told that their concerns were unwarranted (Heifetz 1989, 167–68).

A third, subtler strategy for harnessing medical research to the corporate interest is illustrated in a research program into lead poisoning under the leadership of Robert Kehoe. Kehoe reached his conclusions through the responsible use of scientific methods and published his results in leading journals with no hint of industry censorship. In 1933, he published a series of papers demonstrating a thesis he maintained throughout his long career. He found that significant traces of lead in normal human beings were expected and harmless, that even abnormally high levels in workers' bodies did not necessarily indicate lead poisoning, and that lead would in any case soon descend to normal levels when exposure stopped. As a consequence of these findings, he maintained that the manufacture of tetraethyl lead for gasoline was safe. These findings entailed that lead-using industries needed to employ only minimal safety measures on the job, could continue to manufacture and sell leaded gasoline with impunity, and were well protected from compensation claims.

Kehoe carried out his research from a privileged position. From 1930 to 1958, he was simultaneously a professor at the University of Cincinnati, the director of the industry-funded Kettering Laboratory of Applied Physiology, and medical director of the Ethyl Gasoline Corporation (Graebner 1989, 142). He made no secret of his connections. Did he act independently, or was it his intent to support the lead-based industries? This question is misleading because it directs attention away from the point of this episode: the indistinguishability between industry goals and the goals of scientific medicine represents a coincidence of purpose that does not *require* collusion or control; a moral order has been so constituted as to manufacture this coincidence as a byproduct of their interaction (Foucault 1990, 95).

Kehoe protected his patrons, not by doing shabby science but by carrying out a research program that defined lead poisoning in a manner that was coincident with corporate interests. The corporate effectiveness of this research program is reflected in the persistence of leaded gasoline as a consumer product despite mounting evidence against its use (Hayes 1992, 269–77). At a Senate hearing as late as 1966, Kehoe could still testify as an expert witness against the need for its elimination (Graebner 1989, 142). Although there was some dissent from his views in his later years, real dissent occurred only after 1971 (see Hayes; Matte, Landrigan, and Baker; Needleman, 1992, preface).

Well before the remarkable surge in the growth of occupational medicine, a full integration of medical science and the workplace could be envisioned. In 1908, Dr. Harry Mock, a founding member of the American Association of Industrial Physicians and Surgeons and its second president, marketed his services as an industrial physician to Sears on the basis of the following argument: "Here are 15,000 people, men and women, gathered under one roof in a great industry. *This is a great human laboratory.* Think of the opportunity of making a complete physical examination of all these people. We have been trained to examine sick or supposedly sick people, but here I could study average normal workers and could learn what 'normal' as differentiated from 'sick' really is" (qtd. in Selleck and Whittaker 1962, 61; emphasis in original). Writing nearly two decades later, leading public health physician Alice Hamilton did not share Mock's optimism: "We hope that some way will be found so that . . . animal experiments will precede not follow industrial experiments, and the question [of dangerous work environments] will be treated as one belonging to the public health from the outset, not after its importance has been demonstrated on the bodies of workmen" (qtd. in Heifetz 1989, 173). The mood had changed, though the metaphor had not: it retained its cognitive valence but altered its argumentative aim.

In Mock's metaphoric reconfiguration of workplace, the methodological standards of physics, chemistry, and biology become the defining attributes of occupational medicine: acceptance of a result by the relevant medical community depends solely on its derivation by approved methods. Disease is now defined solely from the point of view of the cognitive technical interest. The nature of this dependence is exemplified by the discovery of the etiology of "radium jaw." In 1924, a New York surgeon diagnosed this condition in a report to a meeting of the American Dental Association; its cause was the poisoning of women who dabbed radioactive paint onto watch-dials, having first sharpened the points of their brushes with their lips and tongues. In 1925, Dr. Frederick Hoffman published an article in the *Journal of the American Medical Association* making the same diagnosis and suggesting the same etiology. In the same year, a confirming study was published in the *Journal of Industrial Hygiene.*

But none of this amounted to medical knowledge. Hoffman is explicit on the point. He interviewed patients, examined X rays, studied conditions at the plant, and drew conclusions concerning etiology based on existing knowledge about the effects of radioactivity. But he is clear that his observations "fall short in dealing with the medical aspects of the cases under observation and, as indicated, a thorough diagnosis fully amplified by all necessary instrumental observations is of the first importance." Only after such an investigation would it be possible to establish "what agent or group of agents was responsible for the pathological conditions in question" (1925, 964).

In the same year, Hoffman's challenge was taken up by Harrison Martland, who established his priority as the discoverer of the cause of "radium jaw" by means of autopsy:

> The cases described in this paper represent practically a hitherto unrecognized form of occupational poisoning. The anemias encountered are for the first time actually proved to be due to the ingestion of radioactive elements with deposition of insoluble, fixed particles in the phagocytic cells of the sinusoids of the reticulo-endothelial system [spaces or tubes through which blood passes in the spleen, lymph nodes, bone marrow, and liver], where they continuously emit irritative rays, which in time produce exhaustion of the adjacent hemapoietic [blood-producing] centers. (Martland, Harrison, Conlon, and Knef 1925, 1775)

Martland argues by analogy from autopsy results in assured cases to the diagnosis of a "Severe Case in Person Still Living." It is this analogical reasoning that enables him to judge that an asymptomatic individual—a "Healthy Case"—is "in grave danger of dying from severe anemia, developing years after [the] original deposition [of radioactive substances] and under conditions in which the *real etiology* must be obscured or never suspected" (1776; my emphasis). The irony inherent in this diagnosis depends entirely on a definition of disease according to the cognitive technical interest. The reliance of diagnosis on autopsy accords with Foucault's observation that medical science privileges not health but death: "a spontaneous *experimental* situation providing access to the very truth of the disease" (1975, 143; my emphasis).

But epistemic closure may be delayed even when post-mortem results are readily available. The etiology of radium jaw turned out to be relatively simple, a short, clear causal chain from occupational hazard to death. Causal chains, however, are seldom so clear-cut; they often involve complex statistical analysis. Reviewing a decade of research on byssinosis, researchers comment that "substantial scientific questions remain about the etiology of byssinosis, the mechanism by which cotton dust acts on the lung, and the relationship between acute and chronic stages of the disease" (Wegman, Levenstein, and Greaves 1983, 191). There were numerous historical, cross-sectional, and prospective studies to serve as "compelling epidemiological evidence of chronic airways obstruction arising from long-term occupational exposure to cotton dust" (Merchant 1983, 138). Nonetheless these studies did not add up to scientific proof (Merchant 1983, 138). The few autopsies that were performed produced results that failed to satisfy the etiological criterion of medical science—a firm statistical correlation between cause and effect: "Post mortem examination of the lungs of cotton workers has shown emphysema and airways disease in varying degrees. [But] one recent study . . . asserted that emphysema was not a consequence of cotton dust but of cigarette smoking" (Wegman, Levenstein, and Greaves 1983, 190).

The motive for delay that certifies occupational medicine as a science is epistemic, not strategic. Nevertheless, epistemic delay can be at odds with worker health; the relentless pursuit of the cognitive technical interest can serve strategic purposes. In 1933, for example, "a negro of 45 came to [the attention of Robert Kehoe and his associates] as a result of an attack of lead intoxication." In addition to the symptoms of lead poisoning, the man had syphilis, swelling and dilation of the aorta, an enlarged heart, and a chronic swelling below the knee in both legs. But "this previously neglected condition was allowed to go untreated until all lead symptoms had disappeared, nor was any treatment given for the lead intoxication, since the subject was in no distress, and since we wished to observe the lead excretion uninfluenced by therapeutic factors" (Kehoe 1933, 324–25).

This is not simply a matter of other times, other standards: a quarter-century later, therapeutic and ameliorative delay remain byproducts of epistemic delay. In 1962, the U.S. Public Health Service and the University of Pittsburgh conducted a study of coke-oven workers and respiratory cancer, an investigation in which researchers discovered that the death rate of "nonwhite" coke workers was 22% above expectation and that their death rate from cancer was double that of other steel workers. A standard textbook states without comment that "while the *tentative* judgment was that this was a *causal* association, it was felt that further information should be collected. *The steel workers were followed for 5 additional years*" (Monson 1990, 237–39; my emphasis).

The delays that high epistemic standards entail are coincident with the economic motive of all industrial health decisions, as the history of beryllium disease also illustrates. The adverse effects of beryllium on worker health were first reported in the 1930s and 1940s. On the basis of this research, a federal standard of worker exposure was set. In 1980, however, a study showed a continued excess of lung cancer in beryllium workers. No one doubted that beryllium was a carcinogen. Nevertheless, a standard textbook published in 1990 comments, "The consistency of findings of an excess occurrence of lung cancer among beryllium workers, while *present*, is not *persuasive*. . . . It does not seem *rational* to spend many millions of dollars on the basis of an excess of 2.4 deaths [per 1,000] from lung cancer" (Monson 1990, 240–44; my emphasis). The text pinpoints the nature of the conflict as "the *cost* in human health vs. the *cost* in dollars" (Monson 1990, 244; my emphasis).

The economic metaphor is not just a metaphor; it is a symptom of a moral order, one in which worker health is fully subordinated to industrial profit: the weighing of these contrasting costs depends, surely, on who is doing the weighing. As editorialized in the *Washington Post*:

> Comparing *relative* costs and benefits is, no doubt, a useful device for deciding among alternative regulations. But in the cotton dust case and all the many others like it, no computation, simple or complex, can absolve Congress of its responsibility for seeing that industries do not pass off the real costs of doing business on either their workers, their communities, or the general taxpayer. (qtd. in Sheehan 1982, 73)

This analysis of a moral order favorable to industrial over worker purposes should not be interpreted as a general bill of indictment against capitalism. We should not see a correlation between capitalism and the risk to worker health and safety; rather, we should see a correlation between the severity of that risk and industrial growth that outpaces the monitoring capacity of governments. In the United States, for example, coal-miner safety compares very favorably with the rest of the world: current miner deaths per million tons of coal in the United States is 0.04; in China, the comparable figure is 1.19 for large state mines, and 9.13 for township and village enterprises—228 times the American figure! The high Chinese figure is the direct consequence of rapid industrialization and government neglect: "The Great Leap Forward led to the neglect of safety work in a rush for production. After a short-lived improvement in the early 1960s, the politics of the Cultural Revolution again led to the neglect of safety work and a higher accident rate" (Wright 2004, 635).

This contrast in industrial safety also holds true when the United States is compared with the former Soviet Union. After the Chernobyl nuclear disaster that killed about thirty persons, a thousand roentgens per hour were recorded near the reactor site twelve days after the accident (fourteen *micro*roentgens is normal; five hundred *micro*roentgens is fatal); 161,000 people were evacuated from a contaminated area that extended to forty-three-hundred square kilometers. In addition, the food supply as far away as England and Finland was seriously compromised. In the case of the Soviet Union, as in the case of China, rapid industrialization coupled with government neglect is the underlying cause of increased risk: "The development of nuclear power in the Soviet Union, and generating power has been put into service as fast as the building of reactors and the long lead times for construction of plants allow" (Shabad 1986, A10). In contrast, in the United States, the Three Mile Island nuclear accident killed no one, and no one was harmed by the small addition to the everyday background radiation ("Chernobyl Accident"; "The Accident at Three Mile Island"). Moreover, after the accident, the American nuclear-power industry amassed an impressive record of strict self-regulation (Rees).

Remaking a Moral Order

The risk to worker health and safety can be reduced only by an alteration in an existing moral order. The routines of society are designed to avoid, not to encourage such alteration. But on occasion, deep underlying conflicts, ordinarily "overlaid by the customs and habits of daily intercourse" manifest themselves "in public episodes of tensional irruption" (Turner 1978, 33, 35). Victor Turner calls these episodes social dramas, his term for real-life sequences of events that share a common underlying dramatic structure. The first act of such a drama, the breach, brings an underlying social conflict vividly to public attention. If not sealed off quickly, this breach will widen into a second act—crisis, a state "coextensive with some dominant cleavage in the widest set of relevant social relations to which the conflicting or antagonistic parties belong" (Turner 1978, 35). Crisis is followed by redressive action: in this third act, society adjudicates rival claims in such arenas of redress as legislatures, regulative bodies, and judiciaries. It is by means of the creation of such social dramas that Habermas's third interest, the cognitive emancipatory, is served, and social orders are remade.

The battle over black lung disease is an example of this process. The story begins in 1968 with a dramatic event, an explosion at Farmington, West Virginia, in which seventy-eight miners were killed. This led almost immediately to a breach in the existing moral order—public anger among the miners followed by political action. The rift between the miners and the industry for which they worked had been widening for some time. The union was no help. Collaboration between the union and the industry over two decades had led only to widespread unemployment through increased mechanization and to severe reductions in pensions and health-and-disability benefits. Breach soon led to crisis: forty thousand miners shut down the coal industry in February, 1969. Led by the newly formed Black Lung Association, they descended on the West Virginia state capital, demanding that Black Lung disease be redefined in their interest.

According to Turner, each phase of social drama is distinct not only in function but another way: "Each has its own speech form and styles, its own rhetoric, its own kind of nonverbal languages and symbolisms" (1978, 43), its own rhetoric of talk and action. On the miners' side, Dr. Isadore Buff made the most of the contrast between his professional status, which gave the miners' cause legitimacy, and a fiery rhetoric aimed at the reform of the existing moral order: "It's up to you men to stand on your own feet and fight, fight, fight, fight, *fight*! The legislature hasn't changed since 1870. They don't really care. But they'd better care right now because you're going to get it. You've got to get it. You've got to get it! Slave or free?" (qtd. in B. Smith 1987, 117).

In contrast, the miners' opponents, the West Virginia Coal Association, appealed for citizen calm in an antirhetoric of objectivity, an antirhetoric aimed at the preservation of the status quo. They deplored the miners' histrionics: "a so-called 'March on the Capitol'; . . . a mock funeral and the flamboyant displaying of so-called diseased lungs to press and legislators" (qtd. in B. Smith 1987, 119). But it was these very histrionics—this very rhetoric of action—that brought about whatever success the miners achieved. As a consequence of their activism, the laws of West Virginia changed in their favor, as did, eventually, the laws of the land. In coal mining in the late 1960s, the miners—not the union, reformers, politicians, or public-health professionals—initiated an insurgency that eventually provided more liberal compensation for Black Lung. The bill that became federal law in 1972 stipulated that no claim to disability could be rejected merely on the basis of a negative X ray; moreover, after fifteen years in the mines, if miners evinced *any* evidence of respiratory or pulmonary impairment, they would be presumed to have an occupational disease (Barth).

Is it correct to call this a redefinition of disease, or would it be more accurate to represent this victory as a triumph of politics over medicine? The latter is the view of the authors of an article on the history of this legislation from 1968 to 1972. In effect, they endorse the opinion of their congressional staff informant that this "was a political decision not a medical one" (qtd. in Fox and Stone 1980, 61–62). At first glance, it would appear that the authors and their informant are entirely correct. Surely, fifteen years in the mines is not the equivalent of a disease-causing agent, a bacillus that would leave an unmistakable spoor in autopsy?

The dismissal of the miners' claims as medical is interested; it is a consequence of the unreflective acceptance of the biomedical model of disease, a model that is entirely a product of the cognitive technical interest and fully coincident with industrial priorities. It is a model, however, whose use is vitiated by the paradox that it militates against worker health. Until phosphorus necrosis, asbestosis, or byssinosis is defined by a medical science driven by the cognitive technical interest, neither match workers, tile workers, nor cotton workers can sicken and die of it; therefore, their complaints are without an object. A disease that is not so defined cannot be prevented or treated. When a person is sick of it, he or she is by definition well; when a person is disabled by it, he or she is by definition not entitled to compensation. So long as disease is defined narrowly according the dictates of a medical science that insists on a strict etiology, so long as attribution requires high statistical correlation shared by a plethora of studies and supported by the biological plausibility of the causal mechanism (Herberg), industry imperatives will be heeded at the expense of worker health. For the needs of workers in the dangerous trades to be served, a breach must be created in the moral order that defines disease solely on the basis of an unequivocally identified pathological derivation. Writing about byssinosis in the *American Journal of Public Health*, medical researchers David H. Wegman, Charles Levenstein, and Ian A. Greaves put the case squarely:

> Occupational health is simply public health applied to the work environment. Such an orientation requires serious efforts aimed at preventing and controlling occupational disease, even in the absence of definitive scientific knowledge. Failure to apply this fundamental approach to the hazards of the workplace has contributed to the continuing political, economic, and scientific controversies over byssinosis, a disabling lung disease of cotton mill workers. Powerful economic interests, reinforced by medical and scientific uncertainty, have delayed and undermined effective measures for the prevention, control of, and compensation for this industrial disease. This, in turn, has resulted in placing the burdens of hazardous employment in the cotton industry on the party least able to bear the weight of public and corporate inaction—the sick worker. (1983, 188)

Wegman, Levenstein, and Greaves resolve the biomedical paradox by supporting a double epistemic standard—higher for medical science considered in isolation from social need, lower for medical science as it is related to actual worker health. This model is certainly more benevolent. But, though it is deeply entrenched, though it informs both our medical and our folk senses of disease (Engel

1977, 130), the biomedical model is inherently problematic as an index of worker health. At its center is a moral order held in place by physicians, medical researchers, and pharmaceutical companies. The only acceptable etiologies terminate in anatomical anomalies causally linked to independent variables supported by statistical methods and capable of isolation and amenable to treatment by standard medical interventions—drugs and surgery. Because it depends for its credibility solely on the evidence of pathological anatomy, the biomedical model marginalizes the known effect of states of mind on physical health and minimalizes more general linkages between disease and ways of life and work.

Alternative models of health and disease—the biopsychosocial and the ecological—take a more liberal view of causation; they broaden the search for causes beyond the physiological. These alternatives are equally plausible, equally "rational." The biopsychosocial model takes into consideration not only "the patient, but the social context in which he lives, and the complementary system devised by society to deal with the disruptive effects of illness, that is, the physician role and the health care system" (Engel 1977, 132). Its goal is a revised medical practice in which grief may be classified as a disease and bad habits tagged as etiological factors. Although its originator does not contemplate a circumstance in which a patient might be not simply an informant concerning symptoms but a player with the power to influence the definition of disease, his concept could easily accommodate such a modification. So modified, the biopsychosocial model would define occupational disease as a consequence of extended negotiations among workers, physicians, public health officials, and industry.

The ecological is another model of health and disease, one in which the divisions between disease-caused and accident-caused disability disappear to the extent that they have a common origin in unsafe working conditions and industrial practices. In this refiguring, unsafe equipment and high levels of coal dust are equally pathogens (in the root sense of the word, they are potential causes of psychological, physiological or physical impairment). In this model, moreover, the line between the workplace and the general environment disappears in medical diagnosis to the extent that it has already disappeared in fact. In the ecological model, we are all victims of the workplace, experimental subjects whose health is being held hostage to the steady deterioration of the environment; we all have a stake and ought to have a voice in responsible industrialization, not only the workers in the dangerous trades. Do we want to know the effect on health and well-being of industrial waste or fast food? Tabulate the rise in pollution-related cancers, in heart disease, in bypass operations.

In the struggle over black-lung legislation, which model of disease was going to prevail, the biomedical, the biopsychosocial, or the ecological? The miners knew that the answer to this question depended on whether employers or employees were doing the defining. Decisions about occupational disease have always been interested; the only difference in this case was that it was the workers' interest that counted.

> If you were an employee in the mines, at work you are covered with dust. It's in your hair, your clothes and your skin. The rims of your eyes are coated with it. It gets between your teeth and you swallow it. You suck so much of it into your lungs that until you die, you never stop spitting up coal. . . . Slowly you notice that you are getting short of breath when you walk up a hill. On the job, you stop more often to catch your breath. Finally, just walking across the room is an effort. . . . Call it miners' asthma, silicosis, coal workers' pneumoconiosis—they are all dust diseases with the same symptoms. (qtd. in Fox and Stone 1980, 46)

In the case of black lung, the miner's discontent created a breach in the social fabric, a prolonged reflective moment in which Habermas's emancipatory interest, embodied in a rhetoric of talk and action, transformed an existing moral order. This transformation altered the understanding of occupational disease, substituting an ecological for a biomedical model. More importantly, the miners' action democratized the process whereby disease is defined. In the case of the biomedical model, the needs of the physicians and their employers were satisfied; in the case of the ecological model, the needs of the miners.

Although the miners actually changed the law as a consequence of their persistent and focused activism and actually and permanently altered the existing moral order, this achievement should not blind us to the limits of their success. From their early insurgency in the coal fields of West Virginia to their successful fight in Washington, the miners would have achieved little were it not for outside funding and the help of sympathetic legislators, physicians, and news media, competent attorneys working for a pittance, VISTA volunteers, and community organizers. It is these professionals who gave the miners' organization its leadership and its focus. Without the deep-seated and general discontent of the miners, it is true, these outsiders would have achieved nothing, but there is no denying either that the power of the workers was mediated at each step by those willing to defy the interests of the class from which they came.

Moreover, this fundamental weakness in worker insurgency was exacerbated by the seriously diverging interests of the miners and of their professional leadership, an underlying fault that led to the eventual collapse of the movement. Throughout, the primary goal of the miners remained adequate disability compensation, while the primary goal of the professional leadership was social change (B. Smith 1987, 161). At first, these goals converged in the joint effort to process successful claims, but the general interest of most of the rank-and-file extended no further than their awards. When the large majority of the claims of the now-retired workers had been adjudicated, they lost interest in the movement (B. Smith 1987, 178–79); furthermore, a new generation of rank-and-file could not be expected to find issues of disability compensation compelling.

Finally, the very extent of the miners' success worked against them. The extraordinary power bestowed by their control over the definition of black-lung disease can be gauged by the insistence even of their congressional supporters that the principles animating black-lung legislation must not be generalized: "When Representative Thompson of Georgia attempted to extend the legislation to cover quarry and textile workers, the House supporters of the bill quickly defeated his proposal" (Fox and Stone 1980, 61). As a consequence of this continuing political anxiety, the possibility of federal relief for occupational disease actually eroded. Peter S. Barth puts it well:

> The program reduced the probability that the federal government will play a significant role in the near future in workers' compensation for occupational disease. Indeed the fact that the efforts to create such a federal presence all failed in the 1970s and 1980s must be attributable in part to the perception that another black lung program, perhaps on a grander scale, would be an expensive blunder. (1987, 284)

In the conclusion of this episode, the last act of this social drama, we see society attempting to minimize the impact of redress and effecting a reintegration of contending parties in a manner most likely to preserve the status quo.

Forms of Redress: The Limits of Citizen Action

Although the miners' victory was only partial, their working conditions, nevertheless, steadily improved throughout the twentieth century. That such improvement occurred under democracy is no accident. While the economic system in place may be irrelevant to worker success, the political system certainly is not. That democracy is not a sufficient condition for worker success is evident from the history of the dangerous trades in nineteenth- and early-twentieth-century America. Nonetheless, it is a necessary condition. This is because democracy provides its citizens with three distinct arenas for effective citizen redress: the legislative, the executive, and the judicial.

Legislators can alter social reality in two ways: through influence and through direct action. In the United States, because legislators are generally driven more by the wishes of their constituencies than by party imperatives, legislator influence can lead to redress in individual cases. Legislative action, however, is, potentially, more far-reaching: to the extent that it is successful, the moral order of society is permanently changed. But change is always a possible threat to the status quo; therefore, increased legislative understanding, rather than enhancing support for change, may easily undermine it, as in the case of the possible extension of black-lung legislation. Legislative action is also vulnerable because it is deliberately unshielded from public and private pressures. This vulnerability creates a continual realignment of political forces. Legislators must respond to the current composition of these forces; their allegiances must constantly shift if these are to remain in line with the shifting priorities of their varying constituencies. Moreover, as a consequence of elections, legislators themselves regularly move in and out of the political process. Finally, legislative action is prospective rather than retrospective: it is constitutionally unsuitable for redress.

Should the miners have chosen executive redress instead, working for a change in the regulative process? Such intervention suffers from none of the deficiencies of legislative redress. As arms of the executive, regulative agencies are durable bureaucracies, far better shielded than legislatures from public and private pressures. Regulative intervention is also more likely to be successful because such redress does not attempt to modify existing moral orders, only to employ them more effectively. A case in point is Love Canal in New York. As a consequence of citizen action, the central portion of the polluted Love Canal site was abandoned, and a portion of the surrounding ring of residences was declared uninhabitable. But the limits of such action are soon reached. The friction built into all bureaucracies ensures that any success will be slow in coming, while the nature of the complaints guarantees that the impact of social action will be largely local (Levine; *History of Disaster*).

Should the miners have chosen the third avenue of redress open to those living in democratic societies, the judicial? Its strengths are considerable: it is better shielded from politics than either legislative or regulative processes; moreover, its impact is general. Even when successful, it is true, judicial redress merely interprets the law. Nevertheless, as a consequence of judicial redress, the law subtly changes, responsive to a new set of interests embodied in altered precedents. The McLibel trial is an example: when a group of London Greenpeace activists accused the corporation of having a deleterious effect on the environment and on public health, McDonald's sued them for libel. Previously, intimidated activists had backed down; but in this case, they chose to stand their ground. After a long trial, the defendants lost: they were found guilty of libel and ordered to pay the corporation £60,000. Nevertheless, the case was clearly a defeat for the plaintiff. McDonald's technical victory was balanced by years of negative publicity, a long-running international story so compelling that by the summer of 2003, the McLibel Web site had just under two hundred million hits ("McLibel

Trial"). Still, activist success was limited. While to some extent, the suit shifted the meaning of libel in British law in favor of defendants, it would be a gross exaggeration to claim that this suit or this shift seriously affected McDonald's business practices or its bottom line.

In democratic societies, workers and citizens have three recourses for redress against moral orders inimical to their health and well-being: the legislative, the executive, and the judicial. In all of these arenas, however, workers' and citizens' victories are subject to limitations that stem in part from the considerable strength and recuperative powers of existing moral orders, in part from the weaknesses of worker and citizen groups themselves. Although industrial organizations compete fiercely against each other, they are nonetheless frequently united in a central aim: creating and sustaining a moral order that emphasizes profitability and marginalizes its social cost. This aim is backed by considerable financial and political power, pursued with ruthless intensity, and defended by means of ceaseless vigilance. Although even a relatively small group of workers or citizens can achieve real gains, nevertheless, difficulties endemic to such groups circumscribe these in the long run. Worker and citizen action are weakened by underfinancing, vitiated by factionalism, and vulnerable to corporate intimidation. The defenders of existing moral orders are well aware of such deficiencies; indeed, they count on them, as Michael Reich's *Toxic Politics* makes clear.

Fortunately, the limits of citizen action do not define the extent of the improvement possible to worker health and safety. This because industries do not—indeed, cannot—define themselves solely in terms of a profitability that marginalizes social costs. Joseph V. Rees's account of the recovery of the nuclear-power industry after the Three Mile Island accident, for example, describes an instance of a moral order that changed as a consequence of action largely internal to the industry. After the accident, as breach threatened to turn to crisis, the industry recognized that each plant was, in effect, hostage to every other: another such incident might well mean the end of nuclear-power generation in the United States. As a result of this collective insight, the industry formed a monitoring and policing organization, the Institute of Nuclear Power Operations (INPO).

Over time, INPO created a moral order that reshaped the nuclear-power industry and made operational safety a priority. By means of a continuing stream of reports that rated the comparative performance of the member plants, INPO succeeded in creating a corporate culture that made potential embarrassment among peers a strong motive for corrective action. When verbal persuasion failed, as in the case of the Peach Bottom, Pennsylvania, nuclear plant, INPO resorted to action, a continuing pressure that led, eventually, to the "retirement" of its top management, giving every other chief executive officer an object lesson in the price of ignoring INPO's standards of performance.

It seems unlikely that this instance of industrial self-regulation is unique; Rees certainly does not think so (1994, 179; 228–29). Just as, in the nineteenth and early twentieth centuries, neglect of worker well-being routinely coincided with industrial purposes, in the late twentieth and the early twenty-first century, a concern for worker well-being may on occasion converge with these purposes. At least, we cannot afford to ignore the possibility that in democratic regimes, worker action and industrial self-interest may work toward the same end, the betterment of worker health and safety.

Part Four

Science Studies: Where Rhetoric Fits In

11 Compatible Insights Between Sociology and Rhetoric

Priority as a Social Norm

In a series of masterful essays, sociologist of science Robert K. Merton developed the thesis that conflicts over priority in science were not, as had been previously assumed, the product of unfortunate personal aberrations. Rather, they constituted a strategic research site for sociological analysis; indeed, they opened a path to the paradox at the heart of the scientific enterprise, the paradox of communal competitiveness. Although the general progress of scientific knowledge relies heavily on the relative subordination of individual efforts to communal goals, the career progress of scientists depends solely on the recognition of their individual efforts at discovery. This paradox and the priority conflicts it engenders were not always a part of scientific activity; they are a historical phenomenon, one whose origin can be traced to England in the middle of the seventeenth century, a time when men deliberated about science and changed their minds concerning the social structure most appropriate to its advance. In the seventeenth century, an ideological shift took pace: the social norms of science changed.

Merton's work exemplifies the need for sociological analysis to identify the structural conditions facilitative of such changes. Nevertheless, norms themselves emerge and gain (or lose) support in contexts in which another sort of analysis seems appropriate—rhetorical analysis. Merton seems to think so himself. In "Priorities in Scientific Discovery," for example, he claims that "property rights in science [consist in] the recognition by others of the scientist's distinctive part in having brought the result into being." As support for this claim, he cites texts employing metaphors in which priority is the thing compared, and property is the basis for comparison: for example, "Ramsay . . . asks Rayleigh's '*permission* to look into atmospheric nitrogen'" (1973, 294–95; my emphasis).

While Merton is sensitive to figurative language in his analysis, he does not place such language at center stage in the transformation of social forces into social action. It is anthropologist Clifford Geertz who turns rhetoric from a category of sociological evidence into a transformative force. To Geertz, it is only through the interaction of social forces and rhetoric that "ideologies transform sentiment into significance and so make it socially available" (1973, 207). Geertz underlines his point that

> the absence of any analytical framework within which to deal with figurative language [has] reduced sociologists to viewing ideologies as elaborate cries of pain. With no notion of how metaphor, analogy, irony, ambiguity, pun, paradox, hyperbole, rhythm, and all the other elements of what we lamely call "style" operate—even, in a majority of cases, with no recognition that these devices are of any importance in casting personal attitudes into public form, sociologists lack the symbolic resources out of which to construct a more incisive formulation [of the operation of ideologies as social forces]. (1973, 209)

It is this transformation by means of figurative language that is the subject of this chapter on priority in science. As we move from text to text, we will see how metaphor opens the door to the redefinition of discovery as a form of private property, a redefinition that culminates in a social drama initiated by Isaac Newton in his dispute with Gottfried Leibniz over the invention of the calculus. This reexamination, this analytical excursion parallel to that of Merton, by making as clear as possible the contrast between rhetorical and sociological analyses, will highlight the value of the former in explaining social change, revealing that changes at the macrolevel have their parallel in changes at the microlevel. Despite rhetoric's transformative role, however, in this chapter, sociology is clearly analytically prior; after all, it is only through sociological analysis that rhetoric's transformative role is discovered.

Bacon's Cooperative Science in *New Atlantis*

The facts concerning the emergence of priority as a social norm in science are not in dispute. At the beginning of the seventeenth century, scientists generally worked independently, disseminating the results of their efforts by means of correspondence and books. From time to time, a scientist might show concern for the order of discovery, but such concern was not an institutional norm. By the end of the century, the situation was radically different: concern for priority in discovery had become just such a norm, with prior journal publication its criterion.

That the story of this ideological transformation begins with a work of fiction necessitates some explanation of the scope of rhetoric: Francis Bacon's *New Atlantis* (1627) is a Utopian novel, not an argument. But for Aristotle, rhetoric is not about argument only: it is about *all* the means of persuasion. *New Atlantis* is a novel with a difference: it makes the claim that a society is best when founded on the advance of science through cooperation; the work persuades not through argument but through its vivid presentation of the benefits of that society. Salomon's House is the intellectual center of the island kingdom of Bensalem. Its goals are lofty: "The knowledge of Causes and secret motions of things, and the enlarging of the bounds of Human Empire, to the effecting of all things impossible" (1937, 480). In pursuit of these goals, the members of Salomon's House perform many of the functions later entrusted to the Royal Society: the members hold meetings, communicate regularly with each other, and publish their results.

At Salomon's House, science is entirely a cooperative activity, pursued according to a firm division of labor: experimenting and the making of inferences are, for example, entrusted to different groups of workers. The science of Salomon's House is, moreover, cooperative in a larger sense; it is an activity entirely free from any taint of nationalism or imperialism: "We maintain a trade . . . only for God's first creature, which was *Light*" (1937, 469). The reference to Deity is apposite because even science is a form of worship: "For the laws of nature are [God's] own laws" (459). In Bensalem, science is an activity totally in harmony with itself, with society at large, and with God's purposes, and it is in no way subordinate to the political purposes of the state (1937, 489). Although the discovery of new knowledge is not a competitive activity—there is no competition anywhere in Bensalem—discoverers are singled out for honor and reward.

In *New Atlantis*, Bacon uses metaphor to underline the high-mindedness of the Utopian vision that is his persuasive vehicle. He does this by varying the distance between the thing compared and the basis of comparison; in doing so, he adjusts "the difficult and elusive relationship between A and B" (Barnfield 1968, 53). For example, throughout *New Atlantis*, Bacon uses light as a metaphor for intellectual and spiritual illumination. In this comparison, there is no cognitive distance between light and the creative act. It is as in *Genesis*: "God said, Let there be light, and there was light." Of course, in this same work, Bacon compares scientific communication to trade and the increase in scientific knowledge to the advance of Empire. In these latter cases, however, he uses these comparisons only to increase the cognitive distance between science and both trade and imperialism: the trade is in light, the Empire human (cf. Jardine 1974, 202–5).

Potential Conflict in Sprat's *History*

Before the founding of the Royal Society in 1662, Englishmen published numerous schemes for the advancement of science whose obvious parent was Salomon's

House (see Jones 1961, 170–76; 317). Thomas Sprat's *History of the Royal Society*, published just five years subsequent to the founding of the society, is in this tradition. It is less the history than the manifesto of that organization; as such, it is a clear embodiment in argument of Baconian myth that sees national and transnational cooperation as the ideal basis for truth-seeking procedures in the natural sciences (1667, 99–100). Sprat notes "how much progress may be made by a form'd and Regular *Assembly*" that coordinates "the joynt force of many men" of all ranks, Englishmen who "*work* and *think* in company, and confer their help to each others *Inventions*" (1667, 28, 39, 427). These cooperative goals are not confined to England. Science prospers "when it becomes the care of united Nations" (3). Cooperation so all-encompassing implies a free exchange of information across national borders: "the benefit of a universal *Correspondence*, and *Communication*" (424).

But in the *History of the Royal Society*, another voice is also heard, a voice at odds with the spirit of cooperation:

> *Invention* is an *Heroic* thing, and plac'd above the reach of a low, and vulgar *Genius*. . . . a thousand difficulties must be contemn'd, with which a mean heart would be broken. . . . some irregularities, and excesses must be granted it, that would hardly be pardon'd by the severe *Rules of Prudence*. (392)

A view of invention so hyperbolic, a stance so obviously motivated by the "desire of glory" (74) is incompatible with Utopian cooperation. Even more jarring to the spirit of cooperation is Sprat's transformation of discovery into a form of capital whose owner needs better protection from society: "While those that add some small matter to things begun, are usually inrich'd thereby; the *Discoverers* themselves have seldom found any other entertainment than contempt and impoverishment. . . . The fruits of their *Studies* are frequently alienated from their Children" (401). In Sprat's myth, in contrast to Bacon's, the heroic and the commercial are indissolubly linked: fellows will "reap the most *solid honor*"; but they "will also receive the strongest assurances, of still retaining the *greatest part of the profit*" (75).

Sprat's other, competitive, voice also reaches beyond national borders: scientific advance is an integral component of England's imperialist destiny. The English character is best suited both to science and to command, indeed, to Empire (113–15, 420). England, therefore, "may justly lay claim, to be the Head of a *Philosophical league*, above all other Countries in *Europe*" (113); London, moreover, is its ideal center: "the head of a *mighty Empire*, the greatest that ever commanded the *Ocean*" (87). Imperialism and science work together. Scientific advance can materially aid "the advancing of *Commerce*, as the best means . . . to enlarge their *Empire*" (408): "if ever the *English* will attain to the *Mastery* of *Commerce*, not only in *discours*, but *reality*: they must begin it by their *labors*, as well as by their *swords*" (423).

When Sprat is praising cooperation, he uses trade and imperialism in a Baconian fashion, as vehicles for metaphors that create cognitive distance between these activities and natural science, an activity whose purpose is "*to increase the Powers of all Mankind, and to free them from the bondage of Errors*, . . . [a] *greater Glory than to enlarge* Empire" (unnumbered epistle dedicatory; emphasis in original):

> By their *naturalizing* Men of all Countries, [the Royal Society has] laid the beginnings of many great advantages for the future. For by this means, they will be able, to settle a *constant Intelligence*, throughout all civil Nations; and make the *Royal Society* the general *Banck*, and Free-port of the World: a policy, which whether it would hold good, in the *Trade* of *England*, I know not: but sure it will in the *Philosophy* [natural science]. (64)

Indeed, in contrast with the imperialist wars of the times, the progress of science is depicted as a war of "all civil Nations" against "*Ignorance*, and *False Opinions*" (57). But when Sprat is waxing patriotic, matters are quite otherwise. England's command over the seas has as its purpose "to bring home matter for *new Sciences*, and to make the same proportion of Discoveries above others, in the *Intellectual* Globe, as they have done in the *Material*" (86).

The aims mirrored in these passages are only potentially in conflict. There is no indication that Sprat—or anyone else at the time—noticed any incompatibility between the cooperative and transnational aims of the Society, on the one hand, and the subordination of science to the national and imperial ambitions of England, on the other. No one, it seems, foresaw the paradox inherent in a *transnational* society founded under *royal* auspices. In all probability, for neither Sprat nor for the founders was this subordination anything more than another argument for the social support of science, one that was, at the same time, a laudable expression of an emerging patriotism. Authors can refuse intellectual engagement with the instabilities of their text, albeit at the price of coherence; but an actual social group simultaneously pursuing cooperation and competition, transnationalism and imperialism, is plainly on a collision course with itself.

Conflicting Norms: The Newton-Leibniz Dispute

In the minds of its founders and of its first Secretary, Henry Oldenburg, the Royal Society embodied the twin virtues of cooperation and transnationalism, the essence of Bacon's vision, the burden of Sprat's argument. In Oldenburg's words: "The object of science [is] of so vast an extent, that it demand[s] the united genius of more than one nation to exhaust the subject" (Birch 1968, 1:317). In 1669 to Christiaan Huygens of Holland, Oldenburg wrote "There is no doubt at all that, if we press onwards at a steady pace, maintaining a frank and regular correspondence for our mutual benefit, we shall in time see considerable progress in every branch of science" (1968, 5:583). In 1667 to René Sluse, the Flemish mathematician, he addressed the following hyberbolic peroration:

> Throw your lot in with us, as the French and Italians have done. We earnestly pray that all your people, all Germany, and the whole of the Netherlands may associate their labors with ours. (1966, 3:537–38)

Throughout his voluminous correspondence, Oldenburg's indefatigable efforts in recruiting and encouraging all scientists, English and Continental, are everywhere apparent (1966, 3:535–38; 1967, 4:419–24; see also 1970, 7:259–60; 336–38). Indeed, so trusted on the Continent was the society under the stewardship of its first secretary—its last Continental secretary—that Oldenburg was able to mediate if not always to resolve disputes, including disputes over priority in discovery, most of which were between Englishmen and their Continental counterparts.

Such disputes arose as a matter of course when the conflicting rhetorical aims in Sprat's *History* became actual social norms, real standards of behavior among the first fellows and correspondents, men who simultaneously worked for and against each other's interests. Even during first years of the Society, problems of scientific ownership appeared (1966, 2:291; 486). And, despite the ready availability of a register of discoveries, priority disputes broke out regularly: between Robert Hooke and Newton; Hooke and Adruen Auzout; Huygens, John Wallis, and Christopher Wren; Huygens and Wallis; Hendrick van Heuraet and William Neile; Huygens and Hooke; Wren and Auzout; Hooke and Gerard Mercator (Westfall 1984, 446–52; Oldenburg 1966, 2:xxii; 1966, 5:374–75; 1975, 10:xxiv, xxvi, 73; Birch 1968, 4:58, 84–86).

Hooke and Newton, Wallis and Wren—these disputing pairs were Englishmen. In the other disputes, however, the conflict was not only between scientists but between England and the Continent. In a letter to Oldenburg, Wallis, the English mathematician, disowned competitive motives on the part of his fellow countrymen: "Whatever may be thought of the French or the Dutch, certainly the English are not thus given to continual pursuit of fame" (Oldenburg 1975,, 10:4). But his more usual views belied these Utopian sentiments:

> Onely I could wish that those of our own Nation; were a little more forward than I find them generally to bee (especially the most considerable) in timely publishing their own Discoveries, & not let strangers [foreigners] reape [th]e glory of what those amongst ourselves are

> [th]e Authors. (Oldenburg 1966, 3:373; 1975 10:41–43, 282; see also 1973, 9:377–78, and Newton *Correspondence*, 1959–77, 1:73; 4:100)

These international disputes reached their climax in the quarrel between Newton and Gottfried Wilhelm von Leibniz over priority in the discovery of the calculus. According to the social theory favored by Victor Turner, this quarrel is a social drama. Insofar as they participate in such a dispute, the actors in a social drama live not by ordinary but by dramatic time and exhibit not everyday but heightened responses (1982, 9–10). Although social dramas have identifiable acts or stages, in a drama as complex and as lengthy as the calculus dispute, it would be futile to try to map these precisely onto the actual sequence of events (1978, 79). This lack of perfect correspondence, however, does not vitiate the theory as an explanation of the text with which we are now concerned, the "Account of the Book entituled *Commercium Epistolicum*," an anonymous summary of the calculus dispute published in 1715 in the *Transactions* of the Royal Society. The "Account" is clearly a form of redress, the stage at which the representatives of a society try to come to terms with the fundamental cleavages that have driven a particular social drama to crisis (1982, 69–70; 78).

The calculus dispute came to a head in the second decade of the eighteenth century. In 1712, under the press of events, the president of the Royal Society convened an international committee to investigate the twenty-year controversy between Newton and Leibniz concerning priority in the invention of the calculus. Within fifty days, they reported in Newton's favor, condemned Leibniz, and issued in evidence a collection of documents, the *Correspondence of John Collins and Others about the Development of Analysis*, usually known by its shortened Latin title, *Commercium Epistolicum*. In 1715, there appeared in the *Philosophical Transactions* of the society a shortened version of the *Commercium*, the "Account."

Any attempt at redress may prove unsuccessful. So it was with the "Account" and with the inquiry that preceded it. The adjudicating body and the documents issued in its wake were so seriously compromised as completely to undermine credibility. The "Account" asserts that "no Man is a Witness in his own Cause. A Judge would be very unjust, and act contrary to the Laws of all Nations, who should admit any Man to be a Witness in his owne Cause" (Hall 1980, 284). But the sole author of the anonymous "Account" was none other than Newton himself. Nor was this his worst. As president of the Royal Society, he had personally selected the committee of inquiry, carefully stage-managed its deliberations, and thoroughly supervised its report.

This brazen behavior had implications far wider than the calculus dispute. From the middle of the sixteenth century, which saw the collapse of the Church as the transnational state, there had been no legitimate governor of the conduct between nations. It is in the century of the Royal Society, the first full century to bear the brunt of this absence, that Hugo Grotius wrote his groundbreaking *On the Law of War and Peace* (1625) "to see whether there were not certain common duties generally felt as binding, if not always practiced, and to set forth an ideal" (Figgis 1960, 246). And it is in the context of this general need for international law that we should see Oldenburg's attempt to legitimate the Royal Society as a means of enforcing common duties in science—what we would call social norms. But Newton's outrageous conduct signaled the utter unraveling of Henry Oldenburg's efforts to transform the Royal Society into a transnational body capable of adjudicating scientific disputes. To this day, no such body exists. A typical contemporary effect of its absence is the continuing priority dispute between American and Russian scientists concerning the discovery of two transuranium elements.

A failure at redressing the dispute between Leibniz and Newton, the "Account" was nevertheless successful at making vivid a vision of science diametrically opposed to that of Salomon's House, one which dramatizes the claim that science is wholly competitive and wholly national in character. The "Account" realizes this vision by inventing two dramatis personae: Newton, who stands for honor, genius, England; Leibniz, who represents immorality, intellectual theft, Continental intrigue. Although the "Account" persuades us neither of Leibniz's guilt nor Newton's innocence, by its twin characterizations it convinces us that the differences between these men are indeed irreconcilable.

By casting the "Account" in the form of a prosecutor's summation before a jury, Newton makes this irreconcilability a given of his case: unrelenting verbal attack becomes not a pleasure to indulge but a duty to discharge. In the following lengthy passage—only length can adequately display his forensic skills—Newton combines historical evidence with sarcasm in a devastating assault on Leibniz's credibility:

> When Mr. *Newton* had received this Letter, he wrote back that all the said four [mathematical] Series had been communicated by him to Mr. *Leibniz*. . . . Whereupon Mr. *Leibniz* desisted from his Claim. Mr. *Newton* also in the same Letter dated *Octob.* 24. 1676. further explained his Methods of Regression, as Mr. *Leibniz* had desired. And Mr. *Leibniz* in his Letter of *June* 21. 1677. desired a further Explication: but soon after, upon reading Mr. *Newton's* Letter a second time, he wrote back *July* 12. 1677. that he now understood what he wanted; and found by his old Papers that he had formerly used one of Mr. *Newton's* Methods of Regression, but in the Example which he had then by chance made use of, there being produced

> nothing elegant, he had, out of his usual Impatience, neglected to use it any further. He had therefore several direct Series, and by consequence a Method of finding them, before he invented and forgot the inverse Method. And if he had searched his old Papers diligently, he might have found this Method there; but having forgot his own Methods he wrote for Mr. *Newton's*. (1980, 279)

Social dramas, as anthropologist Turner discovered working among the Ndembu, tend to widen until they coincide with a society's deepest conflicts; he found that crisis tended to display and redress inscribe these conflicts (1982, 70). It is this tendency that accounts for Newton's view of the priority dispute as an element in the struggle between nations. In English, in an English journal, Newton plainly addresses an audience solely of Englishmen, an audience viewed as both judge and jury (Newton 1980, 309, 310, 313). Of course, to follow the arguments in the "Account," a reader needed a knowledge of both Latin and mathematics; in that sense, Newton's audience was very specialized and not particularly English. But my point here is that the rhetoric of the "Account" creates its own audience, a fiction of Englishmen patriotic to the point of chauvinism.

This tendency for social dramas to widen and deepen accounts also for Newton's conviction that the calculus dispute represents a difference over ultimate issues. Leibniz is wrong not only as a man and a mathematician but in every possible way and at every possible level:

> It must be allowed that these two Gentlemen differ very much in Philosophy [methods of doing science]. The one proceeds upon the Evidence arising from Experiments and Phaenomena, and stops where such Evidence is wanting; the other is taken up with Hypotheses, and propounds them, not by Experiments, but to be believed without Examination.

Indeed, Leibniz is mistaken even about the role of the Almighty in the conduct of the physical world (1980, 314)!

It is another characteristic of social drama that it eventually asserts its independence from its main disputants. In the case of the calculus dispute, the flames of controversy were just as likely to be fanned by others as by Leibniz or Newton; and the quarrel survived Leibniz's death. As it advanced, it deepened; as it deepened, it darkened: the motives of the opposition became automatically suspect, and all attempts at mediation failed. By the time of Newton's "Account," the dispute had been transformed into an irresolvable debate over irreconcilable ideological views, a debate that promised to renew itself in virulence any time two scientists disagreed about priority.

In priority disputes, this tendency of the contestants to lose control made imperative an administratively simple solution that would reduce the frequency of such disputes and contain within the bounds of normal civility those that actually did break out. But before we carry the story of priority to this next stage, the stage at which the institutional response approaches adequacy, we must pause to scrutinize more carefully two assumptions that Newton was careful not to scrutinize at all—that disputes over priority are disputes over intellectual property and that they are national in character.

Redefining Priority as Property

At stake in the calculus dispute was the ownership in perpetuity of this branch of mathematics: exclusively the invention of one Englishman, exclusively the product of the English genius for science. This concern for the scientific ownership that priority affirms is wholly a historical phenomenon. Science is not necessarily conscious of priority; the norms of modern science are not the norms of the ancient Greek or the medieval enterprises of that name. Not until the middle of the seventeenth century did a concern for scientific originality take roughly its present form. To say this is not to deny that earlier scientists might have seen some personal gain in asserting temporal priority for their discoveries; Galileo is a well-known example (Galileo 1957, 232–33, 245).

So persuasive is this idiom that it seems natural today to speak of the *product* of scientific activity as, in some sense, the *exclusive property* of a *person*. Why? In the seventeenth century, at least two notions of scientific discovery were broached. In Huygens's opinion, credit for a discovery "ought to be equally assigned to all those who find out a thing without regard to the time [when it was done], provided that they can state firmly that they have made the discovery without any assistance" (Oldenburg 1968, 5:362). In his controversy with Leibniz over the calculus, Newton asserts his very different point of view: "second Inventors have no Right. The sole Right is in the first Inventor until another finds out the same thing apart. In which case to take away the Right of the first Inventor, and divide it between him and that other, would be an Act of Injustice" (1980, 305; see also 308).

Neither view of scientific discovery is explicable without recourse to the revolution in the law of property that accompanied the rise of capitalism. New forms of enterprise, such as joint-stock companies, required new law; at the same time, the commercialization of society altered the legal status of existing endeavors. From time immemorial, a plow or a book was a form of property, chattel; now, for the first time, the *idea* for a plow or for a book was also a form of property, a very different form, new to the law. Patents and copyrights gave inventors and authors a kind of monopoly. The rights to manufacture and to print could be exclusively owned. As property, they could be bought and sold.

Both Newton and Huygens claim that priority establishes a right in connection with a new sort of intangible property. It is, of course, not a legal right precisely. The right to priority in scientific discovery cannot be licensed, reassigned, or sold, but priority is otherwise precisely analogous to a property right: its proper assertion creates in every scientist the duty to acknowledge another's ownership (Hart 1980, 257–58).

Huygens's view of a common right did not prevail. Its source was in older ideas of property, ideas in the process of being superseded; for example, the feudal notion of the shared rights of tenants to a commons was already endangered and would not survive the eighteenth century. Newton's view, on the other hand, survived and flourished. It clearly rode the crest of the new law of the Stuarts: the exclusive ownership of inventions and literary works guaranteed, in the first case, by the exempting clause of the Statute of Monopolies and, in the second, by the Copyright Act (Jenks 1949, 284–85, 289). Under the sway of such exclusivity, only collaborators and those making a precisely simultaneous independent discovery could share a priority right. This exclusivity is also in line with the English practice of primogeniture, whereby the eldest son inherits all (Pollock and Maitland 1968; Knappen 1964; M. Smith 1928).

This view of priority must be seen in the context of the rise of capitalism, nationalism, and imperialism. Within England, the midcentury saw both agriculture and manufacturing reorganized on competitive lines in accordance with the nascent capitalism on which the industrial revolution would eventually be built. Internationally, in its search for raw materials and markets, England was involved in a bewildering succession of wars and alliances whose only common denominator was international competition, the imperial aim shared by England and the Royal Society. At the end of this period, signalled by the Treaty of Utrecht in 1713, France and the Netherlands, England's colonial rivals, were exhausted by a half-century of warfare. England, on the other hand, was invigorated by fifty years of national and commercial triumph, a vigor manifested very evidently in the rise of English science under the aegis of the Royal Society. It is no accident that in the "Account," Newton's forensic skills are animated by a set of values that were also the basis of the British Empire.

Dealing with Priority Disputes

If the possibility of strife over priority was not continuously to loom, the establishment of an administratively simple means for its determination was imperative. This means had to be persuasive; that is, it had to be acceptable to the community of scientists in the absence of a transnational adjudicating body, an international court of scientific disputes. The means the Royal Society eventually adopted is still in place. Scientific discoveries became and have remained exclusive properties, held by a right analogous to copyright, a right at the same time personal and national; their priority was and is routinely guaranteed by dated journal publication, a procedure that simultaneously fixes the ownership necessary for competitive advantage and guarantees the broadcast of ideas that the imperative to share demands. Because it simultaneously satisfies two potentially conflicting social norms, dated journal publication seems a natural and elegant resolution of the paradox of communal competitiveness (compare Zuckerman and Merton 1973, 465).

Matters were far less clear for scientists at the time. In the first place, in the seventeenth century, the distinction between printing and publishing was very real. Publication was making public; it was the sharing of ideas. Printing was only one means to this end. Writing to Oldenburg in 1665, Auzout states the general view: "I see little difference between printing scientific matters contained in letters and showing these same letters to those learned in these matters who can copy them out when they have them on loan" (Oldenburg 1966, 2:518). Speaking of a scientific discovery, Viscount Brounckner says its author "did then communicate & publish ye same (though not in print) to my self & others" (Oldenburg 1975, 10:291). Giving a lecture was also a means of publication. Wren regards a lecture as "publication enough" (Birch 1968, 1:48). From this point of view, lectures and letters were the equivalent of dated journal publication. Past midcentury, such earlier notions were becoming obsolete; indeed, these last two clarifications are embedded in disputes over priority.

Seventeenth-century scientists were undecided, not only on what publication meant but on whether the broadcast of completed work was the best means of securing priority. Huygens suggested an alternative to full publication: the announcement of work in progress in anagrammatic form (Oldenburg 1968, 5:556). Oldenburg suggested an alternative to anagrams. He was confident that the temporal priority of scientific discoveries could be established by means of the registration stipulated in the founding documents of the Society (Weld 1858, 527). He constantly advertised this option and himself used the register to resolve a dispute (Oldenburg 1966, 2:329; 1966, 3:537; 1967, 4:422; 1968, 5:104, 178; 1975, 10:2, 67). In 1665, he proposed to regularize the process:

> Mr. OLDENBURG made a motion in the name of some member of the society, that when any fellow should have a philosophical notion or invention, not yet made out, and desire, that the same sealed up in a box might be deposited with one of the secretaries, till it could be perfected, and so brought to light, this might be allowed for the better securing inventions to their authors. (Birch 1968, 2:24; see also 212)

But registration was not without problems. The register, of which Oldenburg was so proud, was manifestly incomplete (Sprat 1667, 311; Birch 1968, 3:514; 4:60, 464; Stimson 1948, 66–67). And, even though the Society accepted Oldenburg's proposal, they worried that their fellows might be tempted to register projects far from completion. In an amendment, they insisted that any invention so registered be perfected "after about a year's interval" (Birch 1968, 2:25).

By the seventies, the tide had turned permanently in favor of securing priority by the simple expedient of publication in the newly founded *Philosophical Transactions* (Oldenburg 1975, 10:67). By the last twenty years of the century, the trend toward regular publication became marked. In 1683, the Society passed a resolution to emphasize recordkeeping and putting things in writing (Birch 1968, 4:251). By 1686, there are clear signs of a regular procedure: Denis Papin's paper is read, registered, and published; as are papers by Edmond Halley, Hooke, Gerardus Joannes Vossius, and Sir Theodore De Vaux (Birch 1968, 4:452, 486, 488, 527, 556, 492, 499, 550). The trend toward dated journal publication as a guarantee of priority soon took full hold: by publishing in print accounts of their completed discoveries or inventions, scientists simultaneously shared their results and established their claim to these results as theirs alone. By creating the means by which the ownership rights to its knowledge could be established, the scientific community resolved a dispute and turned the resolution into a social reality.

The Implications of a Social Invention

If the story of priority ended here, it might mislead; it might imply that the consequences of basing the reward system of science on the respective dates of receipt of rival journal articles were uniformly positive. But a focus on priority distorts: it misrepresents the history of science. In any important case, the real question of priority is likely to be beyond settlement. George J. Stigler, a Nobel laureate in economics, asserts:

> When an idea is more than a technical definition or a highly specific analysis, the temporal priority is hopelessly obscure. Who first discovered [such major ideas in economics as] diminishing marginal utility, or diminishing returns, or the quantity theory, or the theory of underemployment equilibrium? I do not know, but it is common knowledge that all such ideas have long histories before they are stated by the men who made them important. (1965, 3)

Thomas Kuhn's classical analysis "Energy Conservation as a Example of Simultaneous Discovery" gives a detailed example of this problem in physics:

> What we see in [the works of these scientists] is not really the simultaneous discovery of energy conservation. Rather it is the rapid and disorderly emergence of the experimental and conceptual elements from which that theory was shortly to be compounded. (1977, 72)

In addition, a focus on priority exaggerates the importance of originality to scientific advance, emphasizing originality to the exclusion of other factors vital to science. The testing of hypotheses, the accumulation of knowledge in a particular field, and the refinement and elaboration of a theory are three important components of scientific advance in which originality is, at best, a minor factor (Stigler 1965).

A focus on priority distorts the history of science in another way. Discoveries as great as the calculus and natural selection were multiple and nearly simultaneous. Had Newton and Charles Darwin never lived, the calculus would have been discovered by the end of the seventeenth, natural selection by the middle of the nineteenth century. How important is the originality that priority certifies when such discoveries are virtually inevitable? Indeed, a concentration on priority entails an emphasis on originality that may actually impede scientific advance:

> Quite aside from the fact that much original work is mistaken, an excessive rate of production of original work may retard scientific progress. . . . [W]hen the rate of output of original work gets too large, theories are not properly aged. They are rejected without extracting their residue of truth, or they are accepted before their content is tidied up and their range of applicability ascertained with tolerable correctness. A cumulative slovenliness results. (Stigler 1965, 14)

Finally, too great a concern with priority distorts the efforts of scientists; it encourages them to show a "concern with recognition" at the expense of a "concern with advancing knowledge" (Merton 1973, 338). In other words, the actual effect of an emphasis on priority may undermine its intended effect, the encouragement of scientific advance. Every scientific paper instantiates the tension created by these opposing tendencies.

Because Merton's conclusions were available to me, my rhetorical analysis follows in the wake of his groundbreaking work. My central insight concerning priority, an insight that organizes this paper, is borrowed from sociology, not earned by rhetoric. But so serendipitous a relationship between sociological and rhetorical analysis is in my view an instance of a more general kinship of two allied disciplines between which a division of labor obtains: sociology still deals with the structural determinants of social conditions, rhetoric with their symbolic interaction in the sphere of social action. In the following and final chapter, I attempt to establish a wider ground for disciplinary cooperation in science studies, to illustrate that the study of science is best served when rhetoric shares the analytical stage with such other disciplines as history, philosophy, and cognitive psychology.

12 Complementary Insights among the Disciplines

The Example of Incommensurability

In the last chapter, although rhetoric performed an explanatory function, it did so in the wake of sociological analysis. Moreover, the role of rhetorical theory in solving the problem of scientific priority was necessarily subsequent to its sociological analysis. Not only did Robert Merton arrive at the essential insights first; he had to arrive at them first. In this chapter, in contrast, rhetorical theory will be shown in an independent role in a cooperative intellectual enterprise: the joint solution of a problem central to science studies. Stasis theory, a component of rhetorical theory, will serve as a general explanatory framework for the solution of a particularly recalcitrant problem in science studies, the incommensurability of paradigms, the central concept in Thomas S. Kuhn's theory of scientific change. In addition, within the framework of stasis theory, rhetorical analysis will join forces with philosophy, history, and cognitive psychology in explicating this central concept.

Stasis Theory

The ancient Greek rhetorical theorist Hermagoras sets down the principles of stasis theory. He stipulates three questions that will determine what is at stake in any courtroom: Has the act been committed? If it has, is it an offense? If it is an offense, are there circumstances that mitigate its seriousness? These are the staseis of fact, definition, and quality. There is also a fourth stasis, that of jurisdiction. This is a question of a different order. Given the nature of our client's alleged offense, we may ask if we are in the appropriate court. If this is civil offense, are we in a civil court? If criminal, are we in a criminal court? (Kennedy 1963, 303–21; Kennedy 1999, 99; Kennedy 1972; Conley 1990, 32–33).

While these are clearly questions designed for legal forums, Hermagoras extended their scope. He applied stasis theory to such courtroom questions as "Did Harmodius and Aristogeiton justly kill the tyrant Hipparchus?" as well as to questions more appropriate to legislative assemblies, questions such as, "Is it right to kill a tyrant?" This broader view of stasis persists in Roman times. In his *Academia*, Cicero uses stasis theory to resolve a philosophical dispute.

According to Richard McKeon, in inventing stasis theory, Hermagoras merely applied Aristotelian philosophical categories to courtroom practice. Specifically, McKeon links stasis theory to the four questions that Aristotle said the sciences must address:

1. Is it?
2. What is it?
3. What properties does it have?
4. What is its proper place in the causal structure of the world? (1949, 5)

Stasis theory—and the philosophical questions that form its basis—have a common origin in Platonic dialectic. According to Hans-Georg Gadamer, that the question is the speech act central to intellectual pursuits is a view that has its origin in Plato in whose dialogues "there is a profound recognition of the *priority of the question* in all knowledge and discourse that reveals something of an object" (1989, 363).

If we begin the history of stasis with Hermagoras, we must agree with contemporary historian of rhetoric Lee S. Hultzén when he endorses "Quintilian's distinction between the first three statuses as rational and the fourth as not a status but a legal question" (103). But if we begin this history with Plato, as I suggest, we can appreciate McKeon's view that Aristotle's last question, that of cause, is not an outlier: "The problem of the *cause* of a process is transformed into the problem of the cause considered and resolved by consensus of the proper body of experts" (1949, 62). On this reading, the fourth stasis is an integral component of stasis theory. For the ancient Greek and Roman rhetoricians, "the proper body of experts" is the appropriate court; by analogy, in matters intellectual, the proper body of experts will be found in the appropriate discipline.

Defined in this way, stasis theory can be fruitfully applied to the problem of incommensurability, a concept so deeply contested that those in science studies can agree neither on its definition nor on its importance. Indeed, the preponderance of authors in a recent collection on the subject deny its existence (Harris), a view so extreme that it can be safely ignored. In what follows, I assume that the first stasis has been satisfied. If it has, the second stasis, that of definition, takes center stage. We want to know what precisely incommensurability is, but the question of quality is equally significant: we need to estimate the importance of incommensurability as a component of scientific change. Perhaps the deep rift in conceptual relations as we move from paradigm to paradigm makes the transition difficult or impossible for scientists and creates a serious hermeneutic problem for historians of science; in contrast, it could be true but trivial that the concepts of a succeeding scientific paradigm are not a perfect match with their predecessor. There is a second reason the stasis of quality is significant: the transition from one paradigm to another must at a minimum have three characteristics: it must be psychologically plausible, it must be communal, and it must be rational in nature. That the transition must be psychologically plausible and communal goes without saying; in the sciences, individuals change their conceptual systems only insofar as change makes sense to them as members of an intellectual community. That this transition must be rational conforms to our sense that, even at its earliest stages, justifications—good reasons—must be given for scientific change.

While in the case of incommensurability, the second and third Hermagorian staseis are operative, the fourth stasis, that of jurisdiction, is importantly prior because, as I will show, the answers to questions of definition and quality are not the property of one discipline but the joint task of four. As a consequence, the solution to the problem of incommensurability exists permanently in an interdisciplinary space. This is not a blended space, such as that of biochemistry, but a space created by the juncture of disciplines, each of which retains its identity. This space is characterized by the absence of routine channels of communication and of the standard professional background consensus, a space in which the dialogue among interested problem solvers is impeded rather than facilitated by prior disciplinary commitments.

That no single discipline can solve the problem of incommensurability is not my contribution to this debate; it is a guiding principle of *The Structure of Scientific Revolutions*. In *Structure*, because conceptual clarification is so central, philosophy of science is patently at work. History of science is also at work because no concept of incommensurability can be credible that does not survive its detailed application to past scientific practice. Because a transition between paradigms is involved, cognitive psychology also plays a role. At times of revolution, when the normal-scientific tradition changes, the scientists' perception of their environment must change radically: they must learn to see the world from a new perspective incompatible with the old (1970, 112). While Kuhn later abandoned his gestalt explanation of the transition, realizing that he had made a category mistake, confusing individual with social psychology, nonetheless, he did not abandon his quest for a psychological component of incommensurability. Nor could he and still claim his explanation was complete.

In *Structure*, rhetoric also has a role, that of explaining the nature of argument in the transition from one disciplinary matrix to another. Rhetoric is essential because, during a period of transition, when the new paradigm is not well established, when good reasons still exist for holding fast to the old, compelling scientific arguments on either side, those ready at hand in the pursuit of normal science, are unavailable (1970, 158; see also 198–204). In later work, Kuhn abandoned this explanatory component entirely, perhaps because he felt that unwelcome conclusions concerning irrationality could be drawn from it. If this is why, he was mistaken: as I will show, while rhetoric remains a necessary explanatory component of incommensurability, no conclusions concerning irrationality need be drawn from its incorporation.

In this chapter, I begin by reconstructing Kuhn's definition of incommensurability. As the issue is conceptual clarification, this reconstruction is a matter primarily for philosophy of science. As my second task, I discuss two aspects of the stasis of quality: the importance of incommensurability to scientific change and the nature of the transition from paradigm to paradigm. The first aspect measures any philosophical analysis of science against the bar of history; the second, because it concerns the explanation of the transition from one paradigm to another, involves both cognitive psychology and rhetoric. These aspects can be addressed, I contend, only in the intellectual space formed by the intersection of the four disciplines concerned: their answer depends crucially on continued dialogue in interdisciplinary space. I conclude this chapter with some reflections on the best way to address problems whose solution disciplinary constraints inhibit, problems such as incommensurability in which the fourth stasis, that of jurisdiction, is interestingly problematic.

The Role of Philosophy of Science

What exactly is Kuhnian incommensurability? To respond is to find oneself firmly within the second Hermagorian stasis, that of definition. It is also to find oneself squarely within the domain of philosophy of science. In Kuhn's early *Copernican Revolution*, a book that preceded *Structure*, incommensurability, though not mentioned, is exemplified at length. Because the revolution is not merely in astronomy but in the way humankind sees itself in relation to the cosmos, Copernican incommensurability is broad in scope:

> As one critic of Dante has put it, in the *Divine Comedy* the "vastest of all themes, the theme of human sin and salvation, is adjusted to the great plan of the universe." Once this adjustment had been achieved, any change in the plan of the universe would inevitably affect the Christian drama of life and death. To move the earth was to break the continuous chain of created being. (1957, 113)

Analogous scientific revolutions are the Darwinian in which humankind's relation to other living creatures is at stake and the Freudian in which humankind's relationship to itself is at issue. In all three cases, the alteration in the perspective of ordinary people was profound: they found themselves living in a different conceptual world. After Copernicus, humankind was no longer at the center of creation; after Darwin, there was no longer an unbridgeable gulf between humankind and other living things; after Freud, humankind found itself at war with itself.

Kuhn did not sustain an interest in this broad view. By the early eighties, he had narrowed his concept of incommensurability to matters scientific and later, even in that arena, somewhat minimized its significance. He insisted that "the claim that two [scientific] theories are incommensurable is more modest than many of its critics have supposed." He averred that were he to rewrite *Structure*, he "would emphasize language change more and the normal/revolutionary distinction less" (2000, 36; 57). He began to conceive of incommensurability in terms of changes in what he called the Lexicon.

In speaking about a Lexicon, Kuhn is talking not about language itself but about what language represents, its underlying concepts. The Lexicon "is a mental module that permits us to *learn to recognize . . . kinds* of physical object (e.g., elements, fields, and forces)" (1993, 315; my emphasis). Scientists learn to identify two sorts of conceptual kinds central to the Lexicon of their discipline: those such as element, compound, and mixture, by contrast; those such as force, mass, and acceleration, by interrelationship.

However they are learned, the terms that designate the conceptual kinds of a developing science have five essential characteristics. First, their meaning and their relationship to other concepts remain only partly articulated: for example, as a consequence of the development of chemistry, our element chlorine and that of Sir Humphrey Davy differ in their internal specification and in their relation to the other elements. Second, despite this partly articulated conceptual core, these terms rigidly designate their objects: for example, over time, a gene can be identified as such by every member of the relevant scientific community. Third, these concepts are projectible; for example, a new planet can be identified by virtue of being of the same *kind* as other planets. Fourth, the borders between these concepts and the kinds they designate are rigidly enforced: there can be no overlaps, except those in which a species falls under a genus. In pre-Einsteinian physics, for example, there can be no overlap between the conceptual kinds of mass and energy; in post-Einsteinian physics, mass and energy are two manifestations of the same conceptual kind. Fifth and finally, the Lexicon itself has a structure, a taxonomy of relationships among kinds. Current chemistry, for example, is unimaginable without the taxonomy of the periodic table; equally, Newtonian mechanics is unimaginable without the law that change of motion is in a straight line and is proportional to the impressed force (2000, 186). It is this Lexical structure, these fundamental relationships among its conceptual kinds, that those practicing a science share and that constitute for them the core of their professional identity.

For those who share a Lexicon of conceptual kinds, the relationships it embodies function as necessary truths (2000, 71). They are not, of course, really necessary truths; in a different Lexicon, they may be falsehoods. This radical change in status occurs, for example, in the shift from the Ptolemaic to the Copernican universe; in the latter, it is no longer a necessary truth that the Earth is a stationary body at the center of the universe. A revolution, then, is a shift from one Lexicon to another, highly incompatible Lexicon of conceptual kinds. After a revolution, the alterations in the scientific Lexicon are so great that it becomes impossible to do science in the old way. After the wave theory of light is firmly in place, no one can do useful scientific work using ray (or selection) theory; after relativity theory is firmly in place, classical physics ceases to be current science. Selection and wave theory and classical and relativity physics are incommensurable.

Despite wide differences in interpretation over the years, I think most commentators would concur that for Kuhn, incommensurability arises out of a serious incompatibility among the conceptual structures of competing paradigms, such as the Newtonian and the Einsteinian. But even were those commentators to agree, first, that there really is such a thing as incommensurability, and, second, that is what Kuhn meant by it, they might still differ concerning its importance in scientific change. It is to this question that I now turn.

The Role of History of Science

The question of the importance of incommensurability to scientific change can be answered only in the interdisciplinary space where philosophy and history of science meet. As Paul Hoyningen-Huene says: "Kuhn's theory is methodologically dependent on the *historiography* of science in that it must allow the latter to furnish it with its particulars" (1993, 12; my emphasis). Kuhn, however, is skeptical about most uses of history as a test for the truth of philosophical analysis: "If prior expectations are not merely unreasonable, the historian committed to them can usually make a case without consciously forcing the data. Accordingly, the historian [and, of course, the philosopher as well] is usually well-advised to set expectations aside before beginning research" (1980, 183). Kuhn would have agreed with Hoyningen-Huene about the importance of history but only so long as historical "data [is] permitted to react back on expectations, make trouble for them, play a role in their transformation" (1980, 182).

It is within this interdisciplinary space between philosophy and history of science that physicist Franz Balsiger and philosopher Alex Burri address the question of incommensurability in their paper "Are Classical Mechanics and Special Relativity Commensurable?" According to them, the question is really three questions:

1. Are the two paradigms compatible syntactically?
2. Are they compatible semantically?
3. Are they compatible in practice?

Their conclusion is that although the two paradigms are commensurable syntactically, that is, although Newtonian can be successfully represented mathematically as a limit condition of Einsteinian mechanics, the two are semantically incommensurable. First, they differ in extension: as one moves from paradigm to paradigm, a term such as *at the same time* shifts in scope; second, they differ in intension: as one moves from paradigm to paradigm, a term such as *mass* shifts in meaning (1990, 161–62). Most important, the two paradigms are incommensurable in practice: "Classical mechanics and special relativity are *self-evidently* incompatible in practice. Not only in engineering, but also in physics we must *decide* in each instance whether relative to the desired use or relative to the desired degree of exactness classical or relativistic methods are required" (1990, 162; my emphasis).

Balsiger and Burri's historical analysis of scientific practice supports Kuhn's philosophical analysis of incommensurability: differences in extension and intension, taken as a whole, taken as a *system* of differences, characterize different paradigms and account for pragmatic incompatibility. Incidentally, this analysis does not support Kuhn's contention that one paradigm must replace another. Balsiger and Burri assert that the criterion for theory choice is less conceptual allegiance than utility. The Newtonian paradigm is still useful, so it is still used but those uses are limited to engineering and scientific applications not at the cutting edge of physics. At the cutting edge of science, one paradigm always replaces another, as Kuhn said.

The work of historian of science Jed Z. Buchwald also supports Kuhn's analysis. Buchwald deals with the case of David Brewster, an English physicist of considerable talent, who never abandoned selection theory, the theory that light is made up of rays rather than waves. Instead, Brewster attempted to integrate the "legitimate" innovations of wave into selection theory. Of his selectionist account of polarization, Buchwald concludes: "It attempts to subvert the vocabulary of the wave theory by appropriating it to selectionist ends, and it does so rather well" (1992, 74). The problem, as Buchwald also points out, is that the price of this accommodation is prohibitively high. It leads to a research program that, although coherent in its own terms, cannot progress because, at each accommodative juncture, a modification has to be made that, however ingenious, is thoroughly ad hoc and, therefore, from the point of view of further research, thoroughly useless. For example, Brewster incorporated into his formulations a component from wave theory, and with it the term *phase*. In so doing, however, he "expropriated a formula [and a term] from the W[ave]-scheme in a way that strips [them] completely of [their] original significance." From the point of view of wave theory, Brewster's use of the term *phase* is, strictly speaking, meaningless, and his formula works only under "degenerate conditions" (1992, 71).

Experimental practice shows that Brewster's attempted accommodations to wave theory lack empirical robustness. As he moves from instrument to instrument, the categories of his taxonomy of polarized light vary unstably:

> The beam-splitter . . . places partial [polarization] *between* elliptic and circular, whereas the special reflection process distinguishes elliptic and circular under a single category *from* partial. This is intensely perplexing because complete and partial are equal siblings [under Brewster's scheme]. Incestuous relationships become possible here. Contrast this with the linear-elliptic-circular categories of W[ave]-polarized [light]. Here beam-splitters place linear before elliptical and circular, while special reflections group the latter two together. Consequently here the beam-splitter enforces a distinction that is also preserved by another device. This is not true in Brewster's taxonomy, where the beam-splitter's distinctions tend to violate the distinctions enforced by special reflections. Brewster's taxonomy is accordingly quite unstable. (1992, 54)

Brewster's failure is especially important because he was a talented physicist who tried to integrate the discoveries of wave into selection theory in a most ingenious matter. He failed precisely because the two taxonomies are conceptually incommensurable at the level of structure and practice, just as Kuhn predicts. One has to choose; one can practice optics *either* according to selection theory *or* according to wave theory.

Balsiger and Burri and Buchwald's cases exemplify a method of which Kuhn approves: the scope and significance of incommensurability can be best explored through an interchange in the interdisciplinary space that the philosophy and the history of science share. In the cases of mechanics and light, we find that physicists cannot practice their science except within the confines of one incommensurable paradigm or another. In these two cases, at least, incommensurability is central to science and to the understanding of scientific change.

The Roles of Psychology and Rhetoric

Another question naturally arises within Hermagoras's third stasis: Is the definition of incommensurability in Kuhn's formulation compatible with a psychologically plausible model of change, one that is both communal and rational? This question is addressed by two philosophers, Paul Thagard and Nancy J. Nersessian, both of whom rely on current paradigms of cognitive psychology. Using artificial intelligence as a starting point, Thagard makes major scientific change cognitively and psychologically plausible. His first step is to establish that concepts, rather than beliefs, are central to any explanation of scientific change. Unlike beliefs, concepts are "complex computational structures" (1991, 102). Most importantly for our purposes, concepts are not isolated but linked in two ways. First, they are connected in networks of part-whole relationships: a whale is a *kind* of mammal; a blow-hole is a *part* of a whale. Second, they are connected by rules: Moby Dick swims *because* he is a whale. These conceptual structures alter over time. From the point of view of paradigm change and its consequence, incommensurability, two sorts of conceptual alteration are crucial: branch jumping and tree switching. Branch jumping is local recategorization; tree switching, its global counterpart. When the Earth becomes just another planet rather than the center of the universe, a branch has been jumped; when biological species are defined not by morphology but by descent, a tree has been switched. Both of these changes—the first, central to the Copernican, the second, central to the Darwinian revolution—lead to major change because both involve significant alterations in *systems* of concepts.

How are these changes effected in a community of scientists? Thagard suggests that a new conceptual network supplants an old one not suddenly but by a gradual weakening of the links within the existing conceptual system and a gradual strengthening of the corresponding links within the new system. This sort of change is psychologically plausible because it involves continuity as well as discontinuity; it assumes that "even revolutionary conceptual change occurs against a background of concepts that have relative stability" (1990, 202). It is also compatible with rationality. Thagard illustrates this compatibility by means of Richard Kirwan's debate with Antoine Lavoisier over the viability of phlogiston theory; it is an example that shows "rational dispute is clearly possible" when paradigm change is at issue (1990, 196). The nature of this debate is central to the demonstration of the rationality of science because it shows that "the major mechanism by which this strengthening [of conceptual links] takes place is scientific *argument*" (1990, 204; my emphasis). But "scientific argument" is not the property of individuals; it is the joint possession of a community of scientists. Thus, for Thagard, conceptual change is both rational and communal.

Thagard's work is usefully supplemented by that of Nersessian. Although fully compatible with that of Thagard, Nersessian's analysis of the "complex computational structures" that are the concepts of science goes further: the finer grain of her taxonomy works better at capturing the structures' internal complexity. For example, Nersessian accounts for the scientific change that was the Maxwellian revolution by means of a taxonomy that involves scientists in simultaneous commitments to four distinctive properties of the electromagnetic field: its ontological status, its function, its mathematical structure, and its causal power. The shift successively from one paradigm to the next—the shift from Michael Faraday to James Clerk Maxwell, then from Maxwell to Konrad Lorentz, and finally from Lorentz to Albert Einstein—is made psychologically plausible because, as with Thagard, it involves continuities as well as discontinuities. An example is the change from Faraday to Maxwell. For Faraday, an electromagnetic field is a substance or a state of the ether; for Maxwell, it is a state of a mechanical ether. For Faraday, the electromagnetic field transmits electrical and magnetic actions continuously through the region surrounding bodies and charges; for Maxwell, it transmits optical as well as electrical and magnetic actions. For Faraday, the mathematical structure of the field is unknown; for Maxwell, it is described by his equations. Finally, for Faraday, the field causes certain electrical and magnetic effects; for Maxwell, it causes all electrical, magnetic, and optical effects as well as radiant heat (Nersessian 1989, 328; see also Nersessian 2003).

While the work of Thagard and Nersessian represents an important response to the problem of transition that incommensurability sets, the shift from paradigm to paradigm cannot be modeled solely by means of the sort of arguments Thagard designates as scientific; it is necessary to resort for a time to broader appeals, that is, to rhetoric. Kuhn states the rationale for this temporary reliance: "The man who embraces a new paradigm at an early stage must often do so *in defiance of the evidence provided by problem-solving.* He must, that is, have *faith* that the new paradigm will succeed with the many large problems that confront it, knowing only that the older paradigm has failed in a few. A decision of that kind can only be made on *faith*" (1970, 158; my emphasis). Though rhetoric is involved, irrationality is not. Kuhn does not say that the shift from an old to a new paradigm is made in defiance of reason but "in defiance of the evidence provided by problem solving." He does not mean that arguments are lacking, only that scientific arguments are. Because those are available only *within* paradigms, they are necessarily absent when paradigm change is at issue.

Rheticus's conversion to Copernicus, the subject of chapter 8, can be reinterpreted in these terms. Rheticus must rely on faith because the principle of parsimony Copernicus professes cannot be realized in the system he propounds; in reality, the system of *De Revolutionibus* is no simpler than the system of Ptolemy's *Almagest.* His shift to Copernicanism is, nevertheless, not without argumentative justification. Rheticus argues as follows: Copernicus was an astronomer following in the footsteps of Ptolemy who, in the course of his work, was forced by his observations and those of other astronomers to question a prevailing Ptolemaic presumption: geocentricity. Such behavior was not eccentric; indeed, it conformed to the principles of inquiry originally championed by Plato and

Aristotle, according to which even assumptions as deeply held as geocentricity must be open to question.

This argument is rhetorical on two counts. First, it is a variation on a central rhetorical argument scheme, dissociation: what looks like discontinuity with the tradition is really, at a deeper level, continuity (Perelman and Olbrechts-Tyteca 1971, 411–59): Copernicus's apparent deviation from a traditional course of inquiry is really a continuation of that course. Secondly, Rheticus's argument is rhetorical because to make its case, it relies upon the authority of the ancients. Although the argument from authority can be fallacious, it need not be; in this case, it is not. Rheticus merely reminds us that most of what we know we know on authority, an authority crystallized in the disciplines we inherit and inhabit. Though Rheticus's argument is rhetorical, it would be a mistake to infer that his conversion by means of it is irrational. Both analyses of scientific change—the conceptual and the rhetorical—are essential parts of its explanation, an explanation that exhibits its psychological plausibility, its avoidance of irrationality, and its communal character.

Rhetoric is sometimes said to imply not only irrationality but also relativism in its more vicious forms, those that deny the existence of a thought-independent world. The rhetorical perspective incorporated in my argument contains no such implication. It is compatible with and would endorse philosopher Helen Longino's view: "One can be a realist in the sense of holding that there is a world independently of our thinking that there is one, without being a scientific realist in the sense of holding that the success of our best theories consists in the world having the exact features attributed to it by those theories" (2002, 141–42).

The Fourth Stasis Revisited

The interdisciplinary space in which I attempt to solve the problem of incommensurability has no current institutional embodiment. That this space does not exist in fact is easily demonstrated by patterns of citation that indicate a lack of intellectual community across the four disciplines involved. Look at the citations in the books and articles used in this chapter. No historian or philosopher of science cites any rhetorician. No philosopher cites Buchwald or Balsiger and Burri. Moreover, these authors do not share a common pool of citations. It is true that Dudley Shapere is cited by Balsiger and Burri and by Nersessian, but the references to Shapere are all marginal to the arguments the citing authors make. Only in the cases of Nersessian and Thagard are common references indicative of a common grounding in cognitive psychology: both cite Carey, Darden, Holland, McCloskey, C. Smith, and E. Smith.

Why does this interdisciplinary space lack a current institutional embodiment? Part of the difficulty in establishing meaningful dialogue among the four disciplines

I employ in this chapter may be traced to an ingrained habit of scholars in the humanities, a strong preference for working alone. This preference accounts for the scarcity of collaborations within the scholarly communities that comprise philosophers, historians, or rhetoricians. In the sciences, in contrast, working alone is not the norm. This is not merely a question of differing social preferences; rather, there is an understanding on the part of scientists that the problems to be addressed should determine working arrangements, not the other way round. The atomic-research facility at Los Alamos provides an example. Of this facility, Robert Oppenheimer, its first director, said:

> I became convinced, as did others, that a major change was called for in the work on the bomb itself. We needed a central laboratory devoted wholly to this purpose, where people could talk freely with each other, where theoretical ideas and experimental findings could affect each other, where the waste and frustration and error of the many compartmentalized experimental studies could be eliminated, where we could begin to come to grips with chemical, metallurgical, engineering, and ordinance problems that had so far received no consideration. (qtd. in Rhodes 1986, 447–48)

The disciplines are necessary components in any viable system dedicated to intellectual advance. Without them, there would be no place for interesting interdisciplinary questions to arise and no one to answer such questions. At the same time, we must recognize that disciplinary allegiances and ingrained habits of solitary work can impede the answers to such questions. To overcome this impediment, we need disciplinary cooperation in both imaginary and in real space. This chapter is an example of its operation the former forum. It revives an ancient idea, Hermagoras's notion of the staseis, in the interest of making a case for rhetoric as an analytical component necessary to the solution of an important intellectual problem in science studies. This particular claim is merely the culmination of the general claim of this book that any study of science that ignores rhetorical analysis is seriously incomplete.

Bibliography
Index

Bibliography

"The Accident at Three Mile Island." United States Nuclear Regulatory Commission. Office of Public Affairs. <http://www.nrc.gov/reading-rm/doc-collections/fact-sheets/3mile-isle.html>.

Allen, Bryce, Jian Qin, and F. W. Lancaster. "Persuasive Communities: A Longitudinal Analysis of References in the *Philosophical Transactions* of the Royal Society, 1666–1990." *Social Studies of Science* 24 (1994): 279–310.

Aristotle. *Aristotle's* Posterior Analytics. Trans. J. Barnes. Oxford: Clarendon, 1975.

———. *The Physics*. Ed. P. H. Wicksteed and F. M. Cornford. 2 vols. Cambridge: Harvard UP, 1929–1934.

Armitage, A. *Copernicus: The Founder of Modern Astronomy*. New York: Barnes, 1962.

Aspect, Alain, P. Grangier, and G. Roger. "Experimental Realization of Einstein-Podolsky-Rosen-Bohm Gedankenexperiment: A New Violation of Bell's Inequalities." *Physical Rev. Letters* 49 (1982a): 91–94.

———. "Experimental Test of Bell's Inequalities Using Time-Varying Analyzers." *Physical Rev. Letters* 49 (1982b): 1804–7.

Bach, Kent, and Robert M. Harnish. *Linguistic Communication and Speech Acts*. Cambridge: MIT Press, 1979.

Bacon, Francis. *The Advancement of Learning*. Ed. G. W. Kitchin. London: Dent, 1962.

———. *Essays, Advancement of Learning, New Atlantis, and Other Pieces*. Ed. Richard Foster Jones. New York: Odyssey, 1937.

———. *Novum Organum*. Trans. R. Ellis and James Spedding. London: Routledge, n.d.

Bale, Anthony. "America's First Compensation Crisis: Conflict over the Value and Meaning of Workplace Injuries under the Employers' Liability System." Rosner and Markowitz 34–52.

Ballentine, L. E., and J. P. Jarrett. "Bell's Theorem: Does Quantum Mechanics Contradict Relativity?" *Amer. Jour. of Physics* 55 (1987): 696–701.

Balsiger, Franz, and Alex Burri. "Sind die Klassiche Mechanik und die Spezielle Relativitätstheorie Kommensurabel?" *Jour. for General Philosophy of Science* 21 (1990): 157–62.

Bambrough, R. "Universals and Family Resemblances." *Wittgenstein: the Philosophical Investigations, a Collection of Critical Essays*. Ed. G. Pitcher. Notre Dame: U of Notre Dame P, 1966. 186–204.

Barnfield, Owen. "Poetic Diction and Legal Fiction." *The Importance of Language.* Ed. Max Black. Ithaca: Cornell UP, 1968. 51–71.

Barth, Peter S. *The Tragedy of Black Lung: Federal Compensation for Occupational Disease.* Kalamazoo, MI: Upjohn, 1987.

Barthes, Roland. *Elements of Semiology.* Trans. Annette Lavers and Colin Smith. New York: Hill, 1968.

———. *S/Z: An Essay.* Trans. Richard Miller. New York: Hill, 1974.

Bazerman, Charles. *Shaping Written Knowledge: The Genre and Activity of the Experimental Article in Science.* Madison: U of Wisconsin P, 1988.

Berkenkotter, Carol, and Thomas N. Huckin. *Genre Knowledge in Disciplinary Communication: Cognition/Culture/Power.* Hillsdale, NJ: Erlbaum, 1995.

Berlin, Isaiah. *Concepts and Categories: Philosophical Essays.* New York: Viking, 1978.

Berry, Arthur. *A Short History of Astronomy from the Earliest Times Through the Nineteenth Century.* New York: Dover, 1961.

Birch, Thomas. *The History of the Royal Society for Improving of Natural Knowledge from Its First Rise.* London: Millar, 1956–1757. Facsim. ed. Ed. A. Rupert Hall and Marie Boas Hall. 4 vols. New York: Johnson, 1968.

Black, Max. *Models and Metaphors: Studies in Language and Philosophy.* Ithaca: Cornell UP, 1962.

Boyle, Robert. "The Origin of Forms and Qualities." *Selected Philosophical Papers.* Ed. M. A. Stewart. Manchester: Manchester UP, 1979. 1–96.

———. *Robert Boyle on Natural Philosophy: An Essay with Selections from His Writings.* Ed. Marie Boas Hall. Bloomington: Indiana UP, 1965.

Brown, Richard Harvey. *Toward a Democratic Science: Scientific Narration and Civic Communication.* New Haven: Yale UP, 1998.

Buchwald, Jed Z. "Kinds and the Wave Theory of Light." *Studies in Hist. and Philosophy of Science* 23 (1992): 39–74.

———. *The Rise of the Wave Theory: Optical Theory and Experiment in the Early Nineteenth Century.* Chicago: U of Chicago P, 1989.

Campbell, John Angus. "Scientific Discovery and Rhetorical Invention: The Path to Darwin's *Origin.*" *The Rhetorical Turn: Invention and Persuasion in the Conduct of Inquiry.* Ed. Herbert Simons. Chicago: U of Chicago P, 1990. 58–90.

Ceccarelli, Leah. *Shaping Science with Rhetoric: The Cases of Dobzhansky, Schrödinger, and Wilson.* Chicago: U of Chicago P, 2001.

Chalmers, Alan. "The Galileo that Feyerabend Missed: An Improved Case Against Method." *The Politics and Rhetoric of Scientific Method: Hist. Studies.* Ed. John A. Schuster and Richard R. Yeo. Dordrecht: Reidel, 1986. 1–31.

"Chernobyl Accident." World Nuclear Assoc. <http://www.world-nuclear.org/info/chernobyl/info7print.htm>.

Chubin, Daryl E., and Edward J. Hackett. *Peerless Science: Peer Review and U.S. Science Policy.* Albany: State U of New York P, 1990.

Close, Frank. *Too Hot to Handle: The Race for Cold Fusion*. London: Allen, 1990.

Cohen, I. Bernard. *Franklin and Newton: An Inquiry into Speculative Newtonian Experimental Science and Franklin's Work in Electricity as an Example Thereof.* Cambridge: Harvard UP, 1966.

———. *Revolution in Science*. Cambridge: Harvard UP, 1985.

Cole, Jonathan R., and Stephen Cole. *Peer Review in the National Science Foundation: Phase Two of a Study*. Washington, DC: Natl. Acad., 1981.

———. *Social Stratification in Science*. Chicago: U of Chicago P, 1973.

Cole, Stephen, Leonard Rubin, and Jonathan R. Cole. *Peer Review in the National Science Foundation: Phase One of a Study*. Washington, DC: Natl. Acad., 1978.

Collins, Harry. *Changing Order: Replication and Induction in Scientific Practice*. Beverly Hills: Sage, 1985.

———. "The Seven Sexes: A Study in the Sociology of a Phenomenon, or the Replication of Experiments in Physics." *Sociology* 9 (1975): 204–24.

Colloquia Copernicana, I. Studia Copernicana, V. Études Sur L'audience de la Théorie Héliocentrique. Wroclaw (Breslau): Polska Akademia Nauk, 1972.

Conley, Thomas M. *Rhetoric in the European Tradition*. New York: Longman, 1990.

Condit, Celeste. *The Meanings of the Gene: Public Debates about Human Heredity*. Madison: U of Wisconsin P, 1999.

Copernicus, Nicholas. *On the Revolutions*. Ed. Jerzy Dobrzycki. Trans. Edward Rosen. London: Macmillan, 1978.

———. *Three Copernican Treatises*. 3rd ed. Trans. and ed. Edward Rosen. New York: Dover, 1959.

Corn, Jacqueline. "Protective Legislation for Coal Miners, 1870–1900: Response to Safety and Health Hazards." Rosner and Markowitz 67–82.

Crick, Francis H. C., J. S. Griffith, and L. E. Orgel. "Codes Without Commas." *Proc. of the Natl. Acad. of Science, U.S.* 43 (1961): 416–21.

Darwin, Charles. *The Autobiography of Charles Darwin: 1809–1882*. Ed. Nora Barlow. London: Collins, 1958.

———. *Charles Darwin's Natural Selection: Being the Second Part of his Big Species Book Written from 1856 to 1858*. Ed. R. C. Stauffer. Cambridge: Cambridge UP, 1975.

———. *Charles Darwin's Notebooks: 1836–1844*. Ed. Paul H. Barrett, Peter J. Gautrey, Sandra Herbert, David Kohn, and Sydney Smith. Ithaca: Cornell UP, 1987a.

———. *The Collected Papers of Charles Darwin*. Ed. Paul H. Barrett. Chicago: U of Chicago P, 1977.

———. *The Correspondence of Charles Darwin*. Ed. Frederick Burkhardt and Sydney Smith. Cambridge: Cambridge UP, 1985–.

———. *Darwin's Ornithological Notes*. Ed. Nora Barlow. *Bull. of the British Museum of Natural Hist.* 2 (1963): 201–78.

———. *Journal of Researches. The Works of Charles Darwin.* Ed. Paul H. Barrett and R. B. Freeman. Vols. 2 and 3. New York: New York UP, 1987b.

———. *The Life and Letters.* Ed. F. Darwin. 2 vols. New York: Basic, 1959.

———. *More Letters.* Ed. F. Darwin. 2 vols. New York: Johnson, 1972.

———. *On the Origin of Species.* 1872. 6th ed. New York: Collier, 1962.

———. *On the Origin of Species.* 1st ed. Cambridge: Harvard UP, 1964.

———. *The Red Notebook of Charles Darwin.* Ed. Sandra Herbert. Ithaca: Cornell UP, 1980.

———. *The Voyage of the Beagle.* Ed. Leonard Engel. New York: Doubleday, 1962.

Davidson, Donald. *Inquiries into Truth and Interpretation.* Oxford: Clarendon, 1984.

———. "Three Varieties of Knowledge." *A. J. Ayer Memorial Essays.* Ed. A. Phillips Griffiths. Vol. 30, Royal Institute of Philosophy lecture series. Cambridge: Cambridge UP, 1991. 153–66.

Davis, Philip J., and Reuben Hersh. "Mathematics and Rhetoric." *Descartes' Dream: The World According to Mathematics.* San Diego: Harcourt, 1986. 57–73.

De Morgan, Augustus. *A Budget of Paradoxes.* Ed. D. E. Smith. 2nd ed. New York: Dover, 1954.

Descartes, René. *Descartes: Philosophical Writings, a Selection.* Ed. Elizabeth Anscombe and Peter Thomas Geach. Edinburgh: Nelson, 1954.

———. *Discourse on Method, Optics, Geometry, and Meteorology.* Trans. Paul J. Olscamp. Indianapolis: Bobbs-Merrill, 1965.

———. *Le Monde, ou Traité, de la Lumière.* Trans. and ed. Michael Sean Mahoney. New York: Abaris, 1979.

———. *Oeuvres de Descartes.* Ed. Charles Adam and Paul Tannery. *Correspondance.* Vol. 2. Paris: Cerf, 1898.

———. *Philosophical Letters.* Trans. and ed. Anthony Kenny. Minneapolis: U of Minnesota P, 1981.

———. *The Philosophical Works.* Trans. Elizabeth S. Haldane and G. R. T. Ross. Vol. 1. Cambridge: UP, 1983.

———. *Principles of Philosophy.* Trans. Valentine Rodger Miller and Reese P. Miller. Dordrecht: Reidel, 1983–84.

Einstein, Albert. *The Principle of Relativity: A Collection of Original Papers on the Special and General Theory of Relativity.* Trans. W. Perrett and G. B. Jeffery. New York: Dover, 1952.

Einstein, Albert, Boris Podolsky, and Nathan Rosen. "Can Quantum-Mechanical Description of Physical Reality Be Considered Complete?" *Physical Rev.* 47 (1935): 777–80.

Eldredge, Niles. Introduction. *Systematics and the Origin of Species.* By Ernst Mayr. 1942. New York: Columbia, 1982. xv–xxxvii.

Engel, George L. "The Need for a New Medical Model: A Challenge for Biomedicine." *Science* 196 (1977): 129–36.

Fahnestock, Jeanne. "Accommodating Science: The Rhetorical Life of Scientific Facts." *Written Communication* 3 (1986): 275–96.

———. *Rhetorical Figures in Science*. New York: Oxford UP, 1999.
Feyerabend, Paul. *Against Method*. 1978. London: Verso, 1978.
———. "Classical Empiricism." *The Methodological Heritage of Newton*. Ed. Robert E. Butts and John W. Davis. Toronto: U of Toronto P, 1970. 150–70.
Figgis, J. N. *Political Thought from Gerson to Grotius: 1414–1625. Seven Studies*. New York: Harper, 1960.
Fitzpatrick, J. W., D. E. Willard, and J. W. Terborgh. "A New Species of Hummingbird from Peru." *Wilson Bull.* 91 (1979): 177–86.
Fleischmann, Martin, and Stanley Pons. "Electrochemically Induced Nuclear Fusion of Deuterium." *Jour. of Electroanalytical Chemistry and Interfacial Electrochemistry* 261 (1989): 301–8; errata, 263 (1990): 187–88.
Foucault, Michel. *The Archeology of Knowledge and Discourse on Language*. Trans. A. M. Sheridan Smith. New York: Pantheon, 1972.
———. *The Birth of the Clinic: An Archeology of Medical Perception*. Trans. A. M. Sheridan Smith. New York: Vintage, 1975.
———. *The History of Sexuality. Vol. 1: An Introduction*. Trans. Robert Hurley. New York: Vintage, 1990.
Fox, Daniel M., and Judith F. Stone. "Black Lung: Miners' Militancy and Medical Uncertainty: 1968–1972." *Bull. of the Hist. of Medicine* 54 (1980): 43–63.
Gadamer, Hans-Georg. *Truth and Method*. Trans. Garret Barden and John Cumming. German, 1965. New York: Crossroad, 1975.
———. *Truth and Method*. Trans. Joel Weinsheimer and Donald G. Marshall. 2nd ed. New York: Crossroad, 1989.
Galilei, Galileo. *Discoveries and Opinions of Galileo*. Ed. Stillman Drake. New York: Doubleday, 1957.
Gaonkar, Dilip Parameshwar. "The Idea of Rhetoric in the Rhetoric of Science." *Southern Communication Jour.* 58 (1993): 258–95. Rpt. in *Rhetorical Hermeneutics: Invention and Interpretation in the Age of Science*. Ed. Alan G. Gross and William M. Keith. Albany: State U of New York P, 1997.
Gaukroger, Stephen. "Descartes' Project for a Mathematical Physics." *Descartes: Philosophy, Mathematics, and Physics*. Ed. Gaukroger. Sussex: Harvester, 1980. 97–140.
Geertz, Clifford. *The Interpretation of Cultures*. New York: Basic, 1973.
Goodman, Nelson. *Ways of Worldmaking*. Indianapolis: Hackett, 1978.
Graebner, William. "Hegemony Through Science: Information Engineering and Lead Toxicology, 1925–1965." Rosner and Markowitz 140–59.
Grant, E. "Late Medieval Thought, Copernicus, and the Scientific Revolution." *Jour. of the Hist. of Ideas* 23 (1962): 197–220.
Greenfield, D. W., and G. S. Glodek. "*Trachelyichthys exilis*, a New Species of Catfish (Pisces: *Auchenipteridae*) from Peru." *Zoology* 72 (1977): 47–58.
Griesemer, James R., and William C. Wimsatt. "Picturing Weismannism: A Case Study of Conceptual Evolution." *What the Philosophy of Biology Is: Essays Dedicated to David Hull*. Ed. Michael Ruse. Boston: Kluwer, 1989. 1–60.

Gross, Alan G. *The Rhetoric of Science.* Cambridge: Harvard UP, 1996.

Gross, Alan G., Joseph E. Harmon, and Michael Reidy. *Communicating Science: The Scientific Article from the Seventeenth Century to the Present.* Oxford: UP, 2002.

Gross, Alan G., and William M. Keith, eds. *Rhetorical Hermeneutics: Invention and Interpretation in the Age of Science.* Albany: State U of New York P, 1997.

Guerlac, Henry. *Newton on the Continent.* Ithaca: Cornell UP, 1981.

Guillemin, Roger, and Roger Burgus. "Hormones of the Hypothalamus." *Scientific American* 227 (Dec. 1972): 23, 33, 134.

Gusfield, Joseph R. *The Culture of Public Problems: Drinking-Driving and the Symbolic Order.* Chicago: U of Chicago P, 1981.

———. "The Literary Rhetoric of Science: Comedy and Pathos in Drinking Driver Research." *Amer. Sociological Rev.* 41 (1976): 16–34.

Haak, Susan. *Defending Science—Within Reason: Between Science and Cynicism.* Amherst, NY: Prometheus, 2003.

Habermas, Jürgen. *Communication and the Evolution of Society.* Trans. Thomas McCarthy. Boston: Beacon, 1979.

———. *Knowledge and Human Interests.* Trans. J. J. Shapiro. Boston: Beacon, 1971.

———. "A Reply to My Critics." Thompson and Held 219–83.

———. *The Theory of Communicative Action. Lifeworld and System: A Critique of Functionalist Reason.* Trans. Thomas McCarthy. Vol. 2. Boston: Beacon, 1987.

———. *The Theory of Communicative Action. Reason and the Rationalization of Society.* Trans. Thomas McCarthy. Vol. 1. Boston: Beacon, 1984.

Hall, A. Rupert. *Philosophers at War: The Quarrel Between Newton and Leibniz.* Cambridge: Cambridge UP, 1980.

Halliday, M. A. K. "On the Language of Physical Science." *Writing Science: Literacy and Discursive Power.* Ed. Halliday and J. R. Martin. Pittsburgh: U of Pittsburgh P, 1993.

Harris, Randy Allen, ed. *Rhetoric and Incommensurability.* West Lafayette: Parlor, 2005.

Hart, H. L. A. "Definition and Theory of Jurisprudence." *Law Rev. Quarterly* 70 (1954): 32–49. Rpt. in *Philosophy of Law.* Ed. Joel Feinberg and Hyman Gross. 2nd ed. Belmont, CA: Wadsworth, 1980. 252–58.

Hayes, Samuel P. "The Role of Values in Science and Policy: The Case of Lead." Needleman 267–83.

Heifetz, Ruth. "Women, Lead, and Reproductive Hazards: Defining a New Risk." Rosner and Markowitz 160–73.

Held, David. *Introduction to Critical Theory.* Berkeley: U of California P, 1980.

Herberg, S. "Validity Aspects of Epidemiological Studies." *Epidemiology of Occupational Health.* Ed. M. Karvonen and M. I. Mikheev. Copenhagen: WHO, 1986.

Herbert, Sandra. "The Place of Man in the Development of Darwin's Theory of Transmutation." *Jour. of the Hist. of Biology* 7 (1974): 217–58; 10 (1977): 155–227.

Herman, Robin. *Fusion: The Search for Endless Energy*. Cambridge: Cambridge UP, 1990.

Heyes, C. M., and H. C. Plotkin. "Replicators and Interactors in Cultural Evolution." *What the Philosophy of Biology Is: Essays Dedicated to David Hull.* Ed. Michael Ruse. Boston: Kluwer, 1989. 139–62.

History of Disaster at Love Canal. Chronology of Events—Part II. <http://ublib.buffalo.edu/libraries/projects/lovecanal/background_gif/records/etf49.html>.

Hoffman, Frederick L. "Radium (Mesothorium) Necrosis." *Jour. of the Amer. Medical Assoc.* 85 (1925): 961–65.

Hoyningen-Huene, Paul. *Reconstructing Scientific Revolutions: Thomas S. Kuhn's Philosophy of Science.* Trans. Alexander T. Levine. Chicago: U of Chicago P, 1993.

Huizenga, John R. *Cold Fusion: The Scientific Fiasco of the Century.* Rochester: U of Rochester P, 1992.

Hultzén, Lee S. "Status in Deliberative Analysis." *The Rhetorical Idiom: Essays in Rhetoric, Oratory, Language, and Drama.* Ed. Donald Bryant. New York: Russell, 1966. 97–123.

Hunt, Bruce J. "Rigorous Discipline: Oliver Heaviside Versus the Mathematicians." *The Literary Structure of Scientific Arguments.* Ed. Peter Dear. Philadelphia: U of Pennsylvania P, 1991.

Jardine, Lisa. *Francis Bacon: Discovery and the Art of Discourse.* Cambridge: Cambridge UP, 1974.

Jarrett, J. P. "Bell's Theorem: A Guide to the Implications." *Philosophical Consequences of Quantum Theory.* Ed. J. T. Cushing and E. McMullin. Notre Dame, IN: U of Notre Dame P, 1989. 60–79.

Jenks, Edward. *A Short History of English Law from the Earliest Times to the End of the Year 1938.* London: Methuen, 1949.

Johnstone, Henry W., Jr. "Can Philosophical Argument Be Valid?" *Bucknell Rev.* 11 (1963): 89–98.

———. *Validity and Rhetoric in Philosophical Argument: An Outlook in Transition.* University Park, PA: Dialogue, 1978.

Jones, Richard Foster. *Ancients and Moderns: A Study of the Rise of the Scientific Movement in Seventeenth-Century England.* New York: Dover, 1961.

Kehoe, Robert A. "Lead Absorption and Excretion in Relation to the Diagnosis of Lead Poisoning." *Jour. of Industrial Hygiene* 15 (1933): 320–40.

———. *The Metabolism of Lead in Man in Health and Disease.* London: McCorquodale, 1961.

———. "On the Normal Absorption and Excretion of Lead. II. Lead Absorption and Lead Excretion in Modern American Life." *Jour. of Industrial Hygiene* 15 (1933): 273–88.

Kehoe, Robert A., Frederick Thamann, and Jacob Cholak. "Lead Absorption and Excretion in Certain Lead Trades." *Jour. of Industrial Hygiene* 15 (1933): 306–19.

Kennedy, George. *The Art of Persuasion in Greece.* Princeton: Princeton UP, 1963.

———. *The Art of Rhetoric in the Roman World.* Princeton: Princeton UP, 1972.

———. *Classical Rhetoric and Its Christian and Secular Tradition from Ancient to Modern Times.* 2nd ed. Chapel Hill: U of North Carolina P, 1999.

Kepler, J. *Mysterium Cosmographicum. The Secret of the Universe.* Trans. A. M. Duncan. New York: Abaris, 1981.

Knappen, M. M. *Constitutional and Legal History of England.* Haden, CT: Archon, 1964.

Koestler, Arthur. *The Sleepwalkers.* New York: Macmillan, 1968.

Köhler, Wolfgang. *Gestalt Psychology.* New York: New Amer., 1947.

Kotelchuck, David. "Asbestos: 'Funeral Dress of Kings'—and Others." Rosner and Markowitz 192–207.

Kuhn, Thomas S. "Afterwords." *World Changes: Thomas Kuhn and the Nature of Science.* Ed. Paul Horwich. Cambridge: MIT, 1993. 224–52.

———. *Black Body Theory and the Quantum Discontinuity, 1894–1912.* Chicago: U of Chicago P, 1978.

———. *The Copernican Revolution: Planetary Astronomy in the Development of Western Thought.* Cambridge: Harvard UP, 1957.

———. *The Essential Tension: Selected Studies in Scientific Tradition and Change.* Chicago: U of Chicago P, 1977.

———. "The Halt and the Blind: Philosophy and the History of Science." *British Jour. of Philosophy of Science* 31 (1980): 181–92.

———. "Reflections on My Critics." Lakatos and Musgrave 231–78.

———. *The Road since Structure: Philosophical Essays, 1970–1993, with an Autobiographical Interview.* Ed. James Conant and John Haugeland. Chicago: U of Chicago P, 2000.

———. *The Structure of Scientific Revolutions.* 2nd ed. Chicago: U of Chicago P, 1970.

Lakatos, Imre, and Alan Musgrave, eds. *Criticism and the Growth of Knowledge.* Cambridge: Cambridge UP, 1970.

Latour, Bruno, and Steve Woolgar. *Laboratory Life: The Construction of Scientific Facts.* Princeton: Princeton UP, 1986.

Leathers, W. S., and Hugh J. Morgan. "The Study of Lead Poisoning in an Enameling Plant." *Jour. of the Amer. Medical Assoc.* 89 (1927): 1107–13.

Leuchtenburg, William E. *Franklin D. Roosevelt and the New Deal: 1932–40.* New York: Harper, 1973.

Levine, Adeline Gordon. *Love Canal: Science, Politics, and People.* Lexington, MA: Heath, 1982.

Lewenstein, Bruce. "From Fax to Facts: Communication in the Cold Fusion Saga." *Social Studies of Science* 25 (1995): 403–36.

Lewontin, Richard C. "The Structure of Evolutionary Genetics." Sober 3–13.

Lindberg, David C. "The Cause of Refraction in Medieval Optics." *British Jour. for the Hist. of Science* 4 (1968): 23–38.

Longino, Helen. *The Fate of Knowledge*. Princeton: Princeton UP, 2002.

Luria, Salvador F. Review of *The Transforming Principle* by Maclyn McCarty. *Scientific American* 254 (Apr. 1986): 24–31.

Lyne, John. "Bio-Rhetorics: Moralizing the Life Sciences." *The Rhetorical Turn: Invention and Persuasion in the Conduct of Inquiry*. Ed. Herbert W. Simons. Chicago: U of Chicago P, 1990. 35–57.

Maestlin, Michael. *Appendix to Kepler's* Mysterium Cosmographicum. Ed. Max Caspar. J. Kepler *Gesammelte Werke*. Vol. 1. Munich: Beck'sche, 1938.

Magnello, Eileen. "The Introduction of Mathematical Statistics into Medical Research: The Roles of Karl Pearson, Major Greenwood, and Austin Bradford Hill." *The Road to Medical Statistics*. Ed. Magnello and Anne Hardy. New York: Rodopi, 2002. 95–123.

Mallove, Eugene F. *Fire from Ice: Searching for the Truth Behind the Cold Fusion Furor*. New York: Wiley, 1991.

Martland, S. Harrison, Philip Conlon, and Joseph P. Knef. "Some Unrecognized Dangers in the Use and Handling of Radioactive Substances with Especial Reference to the Storage of Insoluble Products of Radium and Mesothorium in the Reticulo-Endothelial System." *Jour. of the Amer. Medical Assoc.* 85 (1925): 1769–76.

Matte, Thomas D., Philip J. Landrigan, and Edward L. Baker. "Occupational Lead Exposure." Needleman 155–68.

Mayr, Ernst. *Systematics and the Origin of Species*. New York: Columbia, 1982.

McCarthy, Thomas. *The Critical Theory of Jürgen Habermas*. Cambridge: MIT Press, 1982.

———. Letter to the author. 23 Nov. 1987.

———. "Rationality and Relativism: Habermas' 'Overcoming' of Hermeneutics." Thompson and Held 57–78.

———. "A Theory of Communicative Competence." *Philosophy of the Social Sciences* 3 (1973): 135–56.

McCloskey, Donald N. (Deirdre). *The Rhetoric of Economics*. Madison: U of Wisconsin P, 1985.

McKeon, Richard. "Aristotle and the Origins of Science in the West." *Science and Civilization*. Ed. Robert C. Stauffer. Madison: U of Wisconsin P, 1949. 3–29.

———. *Rhetoric: Essays in Invention and Discovery*. Ed. Mark Backman. Woodbridge, CT: Ox, 1987.

"The McLibel Trial." <www.mcspotlight.org/case/>.

Medawar, Peter D. "Is the Scientific Report Fraudulent? Yes; It Misrepresents Scientific Thought." *Saturday Rev.* 47 (1 Aug. 1964): 42–43.

Merchant, J. A. "Byssinosis: Progress in Prevention." *Amer. Jour. of Public Health* 73 (1983): 137–39.

Merton, Robert M. *The Sociology of Science: Theoretical and Empirical Investigations*. Ed. Norman W. Storer. Chicago: U of Chicago P, 1973.

Miller, Carolyn R. "The Rhetoric of Decision Science, or Herbert A. Simon Says." *The Rhetorical Turn: Invention and Persuasion in the Conduct of Inquiry*. Ed. Herbert W. Simons. Chicago: U of Chicago P, 1990. 162–84.

Moesgaard, K. P. "Copernican Influence on Tycho Brahe." *Colloquia Copernicana 1. Studia Copernicana V. Études Sur L'audience de la Théorie Héliocentrique*. Wroclaw (Breslau): Polska Akademia Nauk, 1972. 31–55.

Monson, Richard R. *Occupational Epidemiology*. 2nd ed. Boca Raton: CRC, 1990.

Montalbo, Thomas. "Winston Churchill: A Study in Oratory." *IEEE Transactions on Professional Communications* 21 (1978): 5–8.

Moss, Jean Dietz. *Novelties in the Heavens: Rhetoric and Science in the Copernican Controversy*. Chicago: U of Chicago P, 1993.

Myers, Greg. "Texts as Knowledge Claims: The Social Construction of Two Biology Articles." *Social Studies of Science* 15 (1985): 593–630.

———. *Writing Biology: Texts in the Social Construction of Scientific Knowledge*. Madison: U of Wisconsin P, 1990.

Needleman, Herbert L., ed. *Human Lead Exposure*. Boca Raton: CRC, 1992.

Nersessian, Nancy J. "Kuhn, Conceptual Change, and Cognitive Science." *Thomas Kuhn*. Ed. Thomas Nickles. New York: Cambridge UP, 2003. 178–211.

———. "Scientific Discovery and Commensurability of Meaning." *Imre Lakatos and Theories of Scientific Change*. Ed. Kostas Gavroglu, Yorgos Goudaroulis, and Pantelis Nicolacopoulos. Dordrecht: Kluwer, 1989. 323–34.

Newton, Isaac. "Account of the Book Entituled *Commercium Epistolicum*." *Philosophical Transactions* 29 (1715): 173–224. Rpt. Hall 263–314.

———. *The Correspondence*. Ed. H. W. Turnbull, J. F. Scott, A. Rupert Hall, and Laura Tilling. 7 vols. Cambridge: Cambridge UP, 1959–1977.

———. *Isaac Newton's Papers and Letters on Natural Philosophy*. Ed. I. Bernard Cohen and Robert E. Schofield. 2nd ed. Cambridge: Harvard UP, 1978.

———. *The Mathematical Papers*. Ed. D. T. Whiteside, M. A. Hoskin, and A. Prag. Vol. 3. Cambridge: Cambridge UP, 1969.

———. *Mathematical Principles of Natural Philosophy and His System of the World*. 1934. Trans. Andrew Motte (1729). Ed. Florian Cajori. 2 vols. Berkeley: U of California P, 1974.

———. *Opticks, or a Treatise of the Reflections, Refractions, Inflections, and Colours of Light*. 1730. 4th ed. 1952. New York: Dover, 1979.

Nirenberg, Marshall W., and J. Heinrich Matthaei. "The Dependence of Cell-Free Protein Synthesis in *E. coli* upon Naturally Occurring or Synthetic Polyribonucleotides." *Proc. of the Natl. Acad. of Sciences* 47 (1961): 1588–602.

Oakeshott, Michael. *Rationalism in Politics and Other Essays*. New York: Basic, 1962.

Oldenburg, Henry. *The Correspondence*. Ed. A. Rupert Hall and Marie Boas Hall. 11 vols. Madison: U of Wisconsin P, 1965–73; London: Mansell, 1975–77.

Patterson, B. D. "Pleistocene Vicariance, Montane Islands, and the Evolutionary Divergence of Some Chipmunks (Genus *Eutamias*)." *Jour. of Mammology* 63 (1982): 387–98.

Patterson, C. "Cladistics." *Evolution Now: A Century after Darwin.* Ed. J. M. Smith. San Francisco: Freeman, 1982. 110–20.

Peirce, Charles Sanders. *Philosophical Writings of Peirce.* Ed. Justus Buchler. New York: Dover, 1955.

Perelman, Chaim, and Lucie Olbrechts-Tyteca. *The New Rhetoric: A Treatise on Argumentation.* Trans. John Wilkinson and Purcell Weaver. Notre Dame, IN: U of Notre Dame P, 1971.

Perutz, Max F. "The Pioneer Defended." *New York Rev. of Books* 21 Dec. 1995: 54–58.

Pinch, Trevor. "Theory Testing in Science—the Case of the Solar Neutrinos: Do Crucial Experiments Tests Theories or Theorists?" *Philosophy of the Social Sciences* 15 (1985): 167–87.

Polanyi, Michael. *Science, Faith, and Society.* Chicago: U of Chicago P, 1964.

Pollock, Frederick, and Frederick William Maitland. *The History of English Law Before the Time of Edward I.* 1895. 2 vols. 2nd ed. Cambridge: Cambridge UP, 1968.

Polya, G. *Of Mathematics and Plausible Reasoning, Vol. 1. Induction and Analogy in Mathematics.* Princeton: Princeton UP, 1954.

Popper, Karl R. *Conjectures and Refutations: The Growth of Scientific Knowledge.* New York: Harper, 1965.

———. *The Logic of Scientific Discovery.* New York: Harper, 1968.

———. "Normal Science and Its Dangers." Lakatos and Musgrave 50–58.

Prelli, Lawrence J. *A Rhetoric of Science: Inventing Scientific Discourse.* Columbia: U of South Carolina P, 1989.

Prowe, L., ed. *Nicolaus Coppernicus.* 1883–84. Vol. 2. Osnbršck: Zeller, 1967.

Putnam, Hilary. *The Many Faces of Realism.* LaSalle: Open, 1987.

———. *Reason, Truth, and History.* Cambridge: Cambridge UP, 1981.

Quine, Willard Van Orman. *From a Logical Point of View: Logico-Philosophical Essays.* New York: Harper, 1961.

———. *Ontological Relativity and Other Essays.* New York: Columbia UP, 1969.

———. "On What There Is." Quine, *From a Logical* 1–19.

———. *The Ways of Paradox and Other Essays.* Cambridge: Harvard UP, 1976.

———. *Word and Object.* Cambridge: Technology P of MIT, 1960.

Rees, Joseph V. *Hostages of Each Other: The Transformation of Nuclear Safety since Three Mile Island.* Chicago: U of Chicago P, 1994.

Reich, Michael. *Toxic Politics: Responding to Chemical Disasters.* Ithaca: Cornell UP, 1991.

Rheticus. *Narratio Prima.* Ed. Henri Hugonnard-Roche and Jean-Pierre Verdet. *Studia Copernica* 20 (1982).

———. *Narratio Prima. Three Copernican Treatises.* Ed. E. Rosen. 2nd ed. New York: Dover, 1959.

Rhodes, Richard. *The Making of the Atomic Bomb.* New York: Simon, 1986.

Ricqlès, A. de, and J. R. Bolt. "Jaw Growth and Tooth Replacement in *Captorhinus aguti* (Reptilia: Captorhinomorpha): A Morphological and Histological Analysis." *Jour. of Vertebrate Paleontology* 3 (1983): 7–24.

Rocke, Alan J. *The Quiet Revolution: Hermann Kolbe and the Science of Organic Chemistry*. Berkeley: U of California P, 1993.

Rohlich, Fritz. "Facing Quantum Mechanical Reality," *Science* 221 (1983): 1251–55.

Roosevelt, Franklin D. *Complete Presidential Press Conferences of Franklin D. Roosevelt*. Vols. 1–2: 1933. New York: DaCapo, 1972.

———. "First Inaugural Address." *Famous Speeches in American History*. Ed. Glenn A. Capp. Indianapolis: Bobbs-Merrill, 1963. 193–98.

Rorty, Richard. *Essays on Heidegger and Others: Philosophical Papers*. Vol. 2. Cambridge: Cambridge UP, 1991.

Rosner, David, and Gerald Markowitz, eds. *Dying for Work: Workers' Safety and Health in Twentieth-Century America*. Bloomington: Indiana UP, 1989.

Rosner, David, and Gerald Markowitz. Introduction. "Workers' Health and Safety—Some Historical Notes." Rosner and Markowitz ix–xx.

———. "Research or Advocacy: Federal Occupational Safety and Health Policies During the New Deal." Rosner and Markowitz 83–102.

———. "Safety and Health as a Class Issue: The Workers' Health Bureau of America During the 1920s." Rosner and Markowitz 53–64.

Ross, D. *Aristotle*. London: Methuen, 1971.

Schlesinger, Arthur M. *The Age of Roosevelt: The Coming of the New Deal*. Boston: Houghton, 1959.

Schrödinger, Erwin. *What Is Life? The Physical Aspect of the Living Cell*. Cambridge: Cambridge UP, 1967.

Schuster, John A. "Cartesian Method as Mythic Speech: A Diachronic and Structural Analysis." *The Politics and Rhetoric of Scientific Method: Historical Studies*. Ed. John A. Schuster and Richard R. Yeo. Dordrecht: Reidel, 1986. 33–95.

Searle, John R. *Speech Acts: An Essay in the Philosophy of Language*. Cambridge: Cambridge UP, 1969.

Selleck, H. B., and A. H. Whittaker. *Occupational Health in America*. Detroit: Wayne State UP, 1962.

Selzer, Jack, ed. *Understanding Scientific Prose*. Madison: U of Wisconsin P, 1993.

Shabad, Theodore. "Atomic Power Gets Priority in Soviet." *New York Times* 29 Apr. 1986: A10.

Sheehan, John J. "Cost-Benefit Analysis: A Technique Gone Awry." *Legal and Ethical Dilemmas in Occupational Health*. Ed. Jeffrey S. Leeb and William N. Rom. Ann Arbor: Ann Arbor Science, 1982. 51–74.

Smith, Barbara E. *Digging Our Own Graves: Coal Miners and the Struggle over Black Lung*. Philadelphia: Temple UP, 1987.

Smith, Munroe. *The Development of European Law*. New York: Columbia UP, 1928.

Snow, John. *Snow on Cholera*. New York: Commonwealth, 1936.

Sober, Elliott, ed. *Conceptual Issues in Evolutionary Biology*. Cambridge: MIT, 1984.

Sokal, R. R., and T. J. Crovello. "The Biological Species Concept: A Critical Evaluation." Sober 541–66.

Solomon, Robert C. "Emotions and Choice." *Explaining Emotions*. Ed. Amélie Oksenberg Rorty. Berkeley: U of California P, 1980. 251–81.

Sprat, Thomas. *History of the Royal-Society of London, for the Improving of Natural Knowledge*. London: Martyn, 1667.

Stigler, George J. "The Nature and Role of Originality in Scientific Progress." *Essays in the History of Economics*. Chicago: U of Chicago P, 1965. 1–15.

Stimson, Dorothy. *Scientists and Amateurs: A History of the Royal Society*. New York: Schuman, 1948.

Sulloway, Frank J. "Darwin's Conversion: The Beagle Voyage and Its Aftermath." *Jour. of the Hist. of Biology* 15 (1982): 325–96.

———. "Darwin's Early Intellectual Development: An Overview of the Beagle Voyage (1831–1836)." *The Darwinian Heritage*. Ed. David Kohn. Princeton: Princeton UP, 1985. 121–54.

Swales, J. M. *Genre Analysis: English and American Research Settings*. Cambridge: Cambridge UP, 1990.

Taubes, Gary. *Bad Science: The Short Life and Weird Times of Cold Fusion*. New York: Random, 1993.

Thagard, Paul. "Concepts and Conceptual Change." *Epistemology and Cognition*. Ed. J. H. Fetzer. Dordrecht: Kluwer, 1991. 101–20.

———. "The Conceptual Structure of the Chemical Revolution." *Philosophy of Science* 57 (1990): 183–209.

Thompson, John B. "Universal Pragmatics." Thompson and Held 116–33.

Thompson, John B., and David Held, eds. *Habermas: The Critical Debates*. Cambridge: MIT Press, 1982.

Turner, Victor. *Dramas, Fields, and Metaphors: Symbolic Action in Human Society*. Ithaca: Cornell UP, 1978.

———. *The Drums of Affliction: A Study of Religious Processes among the Ndembu of Zambia*. Ithaca: Cornell UP, 1981.

———. *The Forest of Symbols: Aspects of Ndembu Ritual*. Ithaca: Cornell UP, 1967.

———. *From Ritual to Theatre: The Human Seriousness of Play*. New York: Performing, 1982.

Virchow, Rudolf. "Typhus and Municipal Sanitation." *Collected Essays on Public Health and Epidemiology*. Ed. L. J. Rather. Vol. 2. New Dehli: Amerind, 1985. 402–35.

Wade, Nicholas. *The Nobel Wars: Two Scientists' Twenty-one-Year Race to Win the World's Most Coveted Research Prize*. New York: Doubleday, 1981.

Walzer, Arthur, and Alan G. Gross. "Positivists, Postmodernists, and Aristotelians and the *Challenger* Disaster." *College English* 56 (1994): 420–33.

Watson, James D., and Francis H. C. Crick. "A Structure for Deoxyribose Nucleic Acid." *Nature* 171 (1953): 737–38.

Weaver, Richard M. "Dialectic and Rhetoric at Dayton, Tennessee." *The Ethics of Rhetoric*. Chicago: Regnery, 1953. 27–54.

Wegman, David H., Charles Levenstein, and Ian A. Greaves. "Byssinosis: A Role for Public Health in the Face of Scientific Uncertainty." *Amer. Jour. of Public Health* 73 (1983): 188–92.

Weld, Charles Richard. *History of the Royal Society, with Memoirs of the Presidents. Compiled from Authentic Documents*. 2 vols. London: Parker, 1858.

Westfall, Richard S. *Never at Rest: A Biography of Isaac Newton*. Cambridge: Cambridge UP, 1984.

Westman, R. S. "The Comet and the Cosmos: Kepler, Maestlin, and the Copernican Hypothesis." *Colloquia Copernicana 1. Studia Copernica 5. Études Sur L'audience de la Théorie Héliocentrique*. Wroclaw (Breslau): Polska Akademia Nauk, 1972). 7–30.

———. "Michael Maestlin's Adoption of the Copernican Theory." *Colloquia Copernicana 4. Studia Copernicana 14. Études Sur L'audience de la Théorie Héliocentrique Copernic et le Développement des Sciences Exactes et Sciences Humaines*. Wroclaw (Breslau): Polska Akademia Nauk, 1975a. 53–63.

———. "The Wittenberg Interpretation of the Copernican Theory." *The Nature of Scientific Discovery: A Symposium Commemorating the Five-Hundredth Anniversity of the Birth of Nicolaus Copernicus*. Ed. O. Gingerich. Washington, DC: Smithsonian, 1975b. 393–429.

Winslow, C. E. A. "Factory Ventilation and Industrial Tuberculosis." *Jour. of the Amer. Medical Assoc.* 85 (1925): 968–73.

Wittgenstein, Ludwig. *Philosophical Investigations*. Trans. G. E. M. Anscome. 3rd ed. New York: Macmillan, 1958.

———. *Preliminary Studies for the "Philosophical Investigations." Generally Known as the Blue and Brown Books*. New York: Harper, 1965.

Woese, Carl R. *Genetic Code: The Molecular Basis for Genetic Expression*. New York: Harper, 1967.

Wright, Tim. "The Political Economy of Coal Mine Disasters in China: Your Rice Bowl or Your Life." *China Quarterly* 179 (2004): 629–46.

Zuckerman, Harriet, and Robert K. Merton. "Institutionalized Patterns of Evaluation in Science." *Robert K. Merton. The Sociology of Science: Theoretical and Empirical Investigations*. Ed. Norman W. Storer. Chicago: U of Chicago P, 1973. 460–96.

Zwerling, Craig. "Salem Sarcoid: The Origins of Beryllium Disease." Rosner and Markowitz 103–18.

Index

Alan G. Gross is a professor of rhetoric at the University of Minnesota. He is the author of *The Rhetoric of Science*, coauthor of *Communicating Science* and *Chaim Perelman*, and coeditor of *Rereading Aristotle's "Rhetoric"* and *Rhetorical Hermeneutics*.